CW00499136

By the Banks of the Lagan

Belfast to Drum Bridge

Ben Simon
2011

The stories told in this book bring to life memories of communities, industries and landscapes by the Lagan from inner city Belfast to the outer suburbs at Drum Bridge. Each of the stories was recorded by the author over cups of tea in people's homes or in cafes, during the period 2008 to 2011. After being transcribed the text was checked with the contributor to allow for any changes, and has been lightly edited, retaining the grammatical errors and repetitions that we all make when speaking, to keep the style and character of the storyteller.

It has only been possible to include stories about a few places along the Lagan, and many locations with interesting histories that deserve to be better known are only briefly described or receive no mention at all. However, it is hoped that this book will encourage further exploration of the heritage of the Lagan Valley and that others will record more memories of life by the riverbanks.

The stories have been grouped into three sections: from the Lagan Weir in Belfast to Stranmillis Weir, around Stranmillis and the first three locks, and from Shaw's Bridge to Drum Bridge. I have given an introduction of 'Lagan lore' at the beginning of each of the three sections to provide an historical context for the stories, and at the end there is a selection of notes and references. These give additional information for anyone who would like to discover more. I would like to thank all the people who kindly agreed to describe their memories of life near the Lagan for this publication. Generous assistance was also provided by Councillor Sara Duncan, Ellison Craig of Stranmillis Residents' Association, Patsy Downey, Dr Jonathan Hamill, Eileen Irvine and Eamon O'Rourke. I would like to thank everybody who contributed comments quoted in the Notes and References section, and I would like to particularly mention the help given by Mrs Frances Weir of Milltown in researching the area around Milltown and Edenderry.

This has been a 'Living Landscapes' project of the Forest of Belfast, undertaken as a contribution to Laganscape, the Lagan Valley Regional Park landscape partnership scheme. I would like to thank both organisations for their encouragement and in particular the assistance provided by Robert Scott and Sue Christie of the Forest of Belfast committee and Brendan O'Connor and Cathy Burns of Laganscape. I am very grateful to Elita Frid for proofreading. Staff of the Public Record Office of Northern Ireland (PRONI), National Museums Northern Ireland, Lisburn Museum, Linenhall Library, Belfast Central Library and Newspaper Library also helped in every way possible. Alex and our daughter Lucia, thank you for your support.

Ben Simon, August 2011

By the Banks of the Lagan

Belfast to Drum Bridge

Ben Simon
2011

THE FOREST OF
BELFAST

ISBN 978-0-9551583-4-6

Designed by Cheah Design.
Printed by Scotprint, Haddington.

Published by Laganscape and Lagan Valley Regional Park,
3 Lock Keeper's Lane, Milltown Road, Belfast. BT8 7XT.
Supported by the National Lottery through the Heritage Lottery Fund.

Contents

Shaw's Bridge and around the river to Drum Bridge

Overleaf: Belfast Harbour from Queen's Bridge with the towboat Sampson pulling a string of barges upriver.

The Lagan through Belfast to Stranmillis

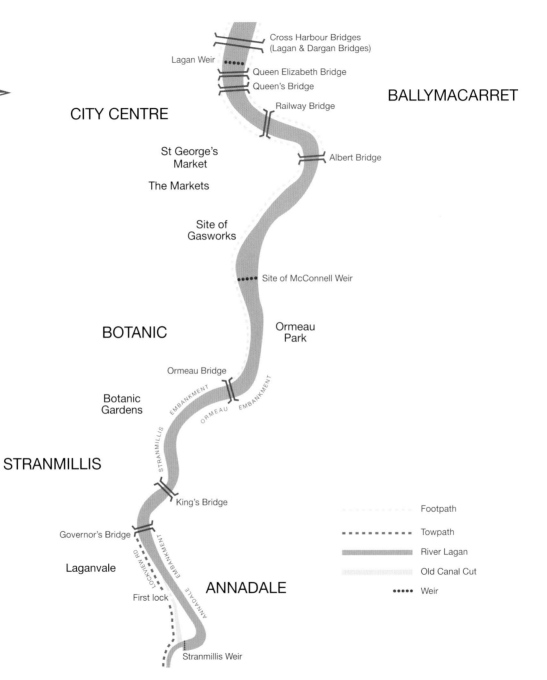

N

CITY CENTRE

BALLYMACARRET

Cross Harbour Bridges
(Lagan & Dargan Bridges)

Lagan Weir

Queen Elizabeth Bridge

Queen's Bridge

Railway Bridge

St George's
Market

The Markets

Albert Bridge

Site of
Gasworks

Site of McConnell Weir

BOTANIC

Ormeau
Park

Ormeau Bridge

Botanic
Gardens

EMBANKMENT

ORMEAU EMBANKMENT

STRANMILLIS

STRANMILLIS

King's Bridge

Governor's Bridge

ANNADALE EMBANKMENT

Laganvale

LOCKVIEW RD

ANNADALE

First lock

Stranmillis Weir

Footpath

Towpath

River Lagan

Old Canal Cut

Weir

The Lagan through Belfast to Stranmillis
Lagan Lore

Belfast, like many cities in the world, is located by a river that has played an important role in shaping its history and character. The origins of the city can be traced back to a settlement founded near the mouth of the River Lagan, at the lowest fording place on the river. As Belfast grew, the ford was replaced by a ferry and by the 1680s a bridge of 21 arches known as the Long Bridge connected Belfast with Ballymacarret.[1]

The Lagan by Belfast was described in a document of 1621 as 'navigable with Barkes and Boates' and as having a salmon fishery where 'the sea doth ebbe and flowe for the space of half a myle up above Belfast'.[2] This was at Stranmillis, where today a weir marks the limit of the tidal estuary.[3] More than a century later, in 1755, fishing in the lower Lagan was still of sufficient importance for Thomas Pottinger of Ballymacarret to try to argue that he had the rights to the river as far as the middle of the channel. Here, he claimed, poles had traditionally been placed to divide the fishery between the owners of the lands on either bank.[4]

Some other glimpses into the ways that residents have valued the river near Belfast can be found in advertisements placed in early issues of the *Belfast News-Letter*. For example a notice of 1764 denounced the person who had shot two swans on the Lagan and offered a reward 'As the Number of Swans on the River adds greatly to the Beauty of the Prospect from the Bridge'. A practical use made of sediments near the mouth of the river is illustrated by a notice of 1788 that announced the sale of 'the benefit of a Shelly Sand Bank, whereby the Town of Belfast is supplied with floor and scouring sand'.[5]

The main docks and quays were developed on the seaward side of the Long Bridge (replaced by the Queen's Bridge in the early 1840s), though with the construction of the Lagan Canal in the mid eighteenth century, barges, known as lighters, travelled upriver from the docks to the first lock at Stranmillis and beyond. There was no towpath before Stranmillis, and it is said that the lighters used a mast and sail on a rising tide to move upriver, empty boats being poled back to the quays.[6] As Belfast developed, an increasing number of bridges were built across the lower Lagan, which must have made it difficult to use sails on the river. In later years, some lighters were motorised, and tugs were used to tow barges to the first lock (in 1899, one resident of Stranmillis wrote to complain about 'the screeching "siren" attached to a steam tug that plies on the Lagan, and startles the neighbourhood at all times of the day and night').[7] Small boats were also used to supply riverside industries including the Gasworks, which opened in 1823 at Cromac, on the County Antrim bank of the river. Initially coal was brought in carts from the quays, but as the works developed, supplies were delivered by lighters which were unloaded by cranes at a wharf at the Gasworks.[8] By the end of the century, lighters on the lower Lagan were also supplying coal to another utility, an electricity generating station situated at East Bridge Street.

The Belfast Amateur Band was perhaps the first organisation that made use of the Lagan near Belfast for recreation. On summer evenings during the 1830s, what were described as musical aquatic excursions were periodically made by boat up the river from the quays, the musicians accompanied by a floating audience, and the riverbanks lined by people listening to the stirring sounds of the band. At one event, a fleet of around 30 boats followed the musicians on the Lagan in what was described as a 'Venetian scene'. On another aquatic excursion, in 1836, the band travelled in a barge decorated with flags pulled by a towboat; passing the Long Bridge, they played the tune 'The Sea', and on reaching Ormeau, then the residence of the second Marquis of Donegall, they struck up 'The Donegall March'. The Marchioness of Donegall watched from a promontory known as the Battery, a riverside ornament armed with cannon, and as the excursion passed, rewarded them by the firing of a salute.[9] The Donegalls, Belfast's foremost family, seemed to have enjoyed these military explosions; on another occasion, the homecoming of Lord Hamilton Chichester, son of the second Marquis, in 1842, it was reported that cannon were discharged from ships in the harbour and 'answered from the battery at Ormeau'.[10]

The Botanic Gardens at Stranmillis, founded in 1828, was linked to the Lagan by a tree-lined walk, and a fete at the garden in 1840 provided an early opportunity for boat and punt races on the river. A newspaper reported that 'the scene on the water was exhilarating, indeed, the river being literally covered with boats, some of which had musical amateurs and their instruments on board'. In one race, however, the oars of two competing boats locked and the confrontation that followed was described as 'a most shameful scene'.[11] A more professional approach to rowing was taken with the formation of the Northern Rowing Club two years later. Its boats were soon competing in style; one account of a race describes the crew of its leading boat *Arrow* as being attired in navy blue jackets and matching caps, the jackets having gilt buttons and a badge displaying the club's initials with a crown and anchor.[12] This was the first of many clubs on the lower Lagan and from these early beginnings, rowing has grown in popularity and continues to be enjoyed to the present day.

The Lagan also provided opportunities for Victorian boating adventures. In January 1867, curious onlookers watched as a young man paddled a single-seat canoe equipped with small masts through the arches of the Queen's Bridge and headed upriver. This trip, however, had to be abandoned at Lisburn because of heavy snowstorms and ice blocking the waterway. On a subsequent expedition, two canoeists succeeded in journeying to the end of the Lagan Canal and into Lough Neagh. The difficulties that had to be overcome on this journey included carrying canoes over locks, allaying the suspicions of the constabulary at Moira, surviving an exploding camping stove, weathering a squall on the lough and being woken at night by a heifer stumbling onto a tent at Coney Island.[13]

The riverside grounds of the second Marquis of Donegall at Ormeau were acquired by Belfast Corporation to create Belfast's first park which opened in 1871 and, together with the Botanic Gardens

on the opposite bank, provided much-needed public space by the Lagan. However, with the rapid growth of Belfast during the latter part of the nineteenth century, large sections of the riverbanks were used for industries including brickworks which lined the riverbank from Ormeau as far upriver as Annadale. At Stranmillis there were rubbish dumps and ash pits by the river, and the development of Belfast also gave rise to a problem that some politely referred to as the 'Blackstaff nuisance'.

George Benn, author of *A History of the Town of Belfast*, published in 1877, described how, in the previous century, the townsfolk used to fish in the clear water of the Lagan and bathe in the Blackstaff, a tributary which flows from west Belfast to the Lagan at lower Ormeau. George Benn also added a footnote in his book to advise readers that it was 'actual fact' that the Blackstaff had previously been pure enough for bathing.[14] The reason why readers might have doubted the story was that at the time that Benn was writing, the streams near Belfast were so polluted that it would have been unthinkable to immerse oneself in the water. The smell from the Blackstaff gave rise to concern in the 1840s and to increasingly angry complaints in the following decades. A ballad published in 1874 gave, in its two short verses, a description of the problem.[15]

> The Blackstaff flows by noisome spots
>   A dark and deadly river,
> Foul stenches its forget-me-nots
>   Which taint the air for ever.
> It gushes, glides, it slips, it slides,
>   And mocks each poor endeavour;
> For all we write and all be talk,
>   It reeks, and reeks for ever.

> It reeks with all its might and main,
>   Of death and plague the brewer,
> With here and there some nasty drain,
>   And here and there a sewer.
> By fetid bank, impure and rank,
>   It swirls, a loathsome river;
> Its breath is strong, though we be weak,
>   And death it brews for ever.

It was decided to culvert the Blackstaff, hiding the smell and pollution underground,[16] though the stream continued to discharge sewage and industrial waste into the Lagan. The summer of 1885 was described by one resident as 'a best on record, as the smell was more offensive than ever, and not content with keeping in its own quarters, penetrated right up to South Parade and Ballynafeigh on the one side, and to The Plains [Malone] on the other'.[17] Eventually, following the holding of public meetings, setting up of a Lagan Pollution Committee and convening of special meetings of the town council, improvements were made to the sewer system.[18] In addition to being polluted, the lower reach of the Lagan was very shallow, with extensive mud banks exposed at low tide. In the late 1880s, the Lagan was dredged as far upstream as Stranmillis, and the first lock was rebuilt with a lower sill. This task was primarily undertaken to ensure that lighters could move upriver and enter the Lagan Canal at all states of the tide, though this also enhanced the visual appearance of the river.[19]

The Lagan in 1929, during the construction of the sloping stone banks as part of the Boulevards scheme. The photograph was taken from Botanic with Ormeau Bridge in the distance.

In 1924, an act was passed to enable Belfast Corporation to improve the amenity of the tidal Lagan between lower Ormeau and Stranmillis.[20] The McConnell Weir (named after Alderman Sir Thomas McConnell, who had promoted the project) was constructed downstream of the Ormeau Bridge to impound the river water between the weir and Stranmillis. It opened in 1937 and incorporated an electrically operated lock to enable boats and canal lighters to pass.[21] Roads entitled the Lagan Boulevard were constructed along the riverbanks at Ormeau, Annadale and Stranmillis, and in the following years strips of open space were created along the riverside. These green corridors connected the Botanic Gardens and Ormeau Park with the countryside to the south of the city and also helped to mask industries including brickworks that were still operating at Annadale.

In the post-war years there was a reduction in river traffic and by the mid-1950s, less than 20 years after the lock at the McConnell Weir was opened, it had become almost completely unused. Commercial activity on the river ceased with the ending of barge traffic carrying coal to the Gasworks in the early 1960s.[22] Employment in the docks and shipyards at the mouth of the Lagan was also decreasing and, in the following decades, the Belfast markets around Oxford Street and many of the factories along the river either closed or relocated away from the inner city. Communities that had grown in partnership with these industries at places like the Markets, Short Strand, lower Ormeau and Laganvale also declined. Poor water quality, dumping and the buildup of polluted mud banks in the Lagan added to the sense of dereliction. The riverside Boulevards lost their appeal and became known by the more functional name of embankments. Belfast had turned its back on the Lagan and its heritage.

Black guillemot on a pier of Queen's Bridge, 2002.

Around this period, many other cities in Britain and Ireland with a maritime history similarly had redundant and decaying docklands and disused waterways, and there was a move to regenerate river frontages and reconnect cities with their rivers and harbours. The Laganside Concept Plan, published in 1987, aimed to enhance the river through Belfast together with adjacent, old industrial and commercial sites, including redundant markets and the Gasworks site.[23] A new weir, quays and riverside walkways were built, the river was cleared of dumped material and efforts were made to improve water quality. A cross-harbour road and rail link were constructed to take traffic away from the old Lagan crossings and free up roads around the river and city centre. This enhanced environment was designed to encourage investment in the area and re-establish the city's links with the Lagan.

Today, just 25 years since the implementation of the Laganside project, the results have been impressive and the lower reaches of the Lagan are now a focal point for the city. Iconic buildings, such as the former Gasworks offices and clock tower, St George's Market and the Albert Clock, have been beautifully restored. Disused riverside lands and redundant quays have been redeveloped for business and housing and incorporate public spaces and two major public venues, the Waterfront Hall and Odyssey Centre. The improving condition of the Lagan is encouraging wildlife to return, and greater use is also being made of the Lagan by rowers. In recent years, there have also been boat tours, duck boat trips and the MV *Confiance*, a restored barge moored by the Waterfront Hall, is now open with displays of maritime heritage.

A cow came up our hall one time
Rose Ann Smyth

I grew up in Annette Street and it was beautiful, lovely, I loved it and all the markets facing me. The three fruit markets, the two wholesale markets and the people used to go whenever the fruit was brought in, in cases. The orange boxes and the apple boxes and all. Now the banana place was away round by the docks, the banana market. But this was the fruit market – the apples and oranges and all, plums and damsons and grapes and that kind of stuff. The next one to that was the butter and egg market, that was a great market.[24]

You could get everything. Watercress, marvellous, you couldn't get better than watercress for iron, it was the best thing in the world for you. A penny bunch you got. Now it's one pound and it's half dead, we wouldn't have used it! We got garlic and the garlic would have put you out of the house, one wee spud of it and you had to wash your mouth out before you went out with your boys with the taste of the garlic. Now you could open up the whole lot and there's no smell.

That's the thing I always remember as a child. Wakening up in the morning and the country carts they were coming after picking the fruit and packing it and it was all in bundles. Bundles of parsley, bundles of celery, bundles of beetroot. Don't forget every different cart had different things in it. All turnips, beautiful, you could eat a turnip; now you wouldn't eat the thing but your mammy would have been cutting it and you would have half of it ate, do you understand? So nice, the turnips. The whole cart full of turnips, the next one full of cabbage, white cabbage and what do you call it, pamphrey. The different carts. And then there would be cauliflowers, all gorgeous. Then the leeks and the parsley and the celery. You smelt everything, there was a smell of everything, you understand. They would have been waiting all on the hill coming at the market, coming down to the market at Oxford Street. Down the hill at the Albert Bridge. And they would have stopped at the potato market waiting on it opening. The country people came in with their stuff and put it all on the ground. Lovely neat piles. And then the shop people would come in and buy stuff and if they didn't get it all sold then we could have come in and we bought. Do you understand me now?

They would have been there from about half five or so and there was an eating house on the corner of Oxford Street and you would have smelt the bacon and eggs, you understand. Soda bread, the potato bread, pancakes, black pudding and bacon and sausages. Beautiful. And then for the afternoon they would have had soup or a pot of stew on, you understand. And there used to be an old man lived up Hartrick's Court. Hartrick's Court was six, no eight houses, at Annette Street, you went up an entry and there was four houses went to this toilet and four houses went to that toilet, right? And Dicky Bultitude – that's what you called him, that was his name – and he used to come out with a wee thing, it was like an apple box he had with wee wheels on it and two sticks, handles on it. And in that he had a bottle of whiskey and a bottle of wine or something. But that was covered, you see, and he went and with the one cup, now, the one glass, done everybody. And the country men after coming in on a cold morning they got their drink. It was a great place.

In Hartrick's Court there was just a room downstairs; no coal hole, no water, no scullery, no nothing, and one upstairs for a bedroom, and the bedroom was just mattresses filled with straw or something, and threw down for the children to lie there head to toe. And behind the door was the water tap. Behind the front door. No sink, nothing. A bucket, you went to your toilet up the stairs with a bucket, right? Out on the landing. They had lines, lines out there in Hartrick's Court, lines for their washing. That was it.

I grew up in 17 Annette Street and my grandparents lived at number 13, but oh, we were swanks, we had a parlour you see, and a piano. Oh, a piano, we had a piano from no age and my mammy was great at the guitar and the wee squeeze box. You come up the hall, the hall stand, and there was a big curtain there to hide the stairs, and on the stairs you had your carpet and you had your brass stair rods and they had to be cleaned and all. The mats in the hall and you had to watch you didn't polish too much because if you did you would slide away up the hall you see. So you had to put the mats down after you washed. You put the mats down then you just polished round. And your knocker and the number of your door, your handle of the door, and every day the window sill was washed. Every day, not only us, the whole street. And all brush down the front and all after you beat your mats up against the wall, then you came out and you threw the disinfectant around the front because the kids had the whooping cough, you see, and they went to the grating and spit in it. And they went up to the Gasworks, to be put into the Gasworks. When you went into the Gasworks you looked in. The man he would have said, 'Right', and he would have come and took whatever child had the whooping cough and he brought them up and you went up with them. And you had to wait until they coughed, you see. And, 'Right daughter, bring her back tomorrow.' Three times. And the Murdocks used to come round with the wee donkey. They were great people at the markets, a great family. They were in delft and fowl and shops and all. He would come round with the donkey and when he come round with the donkey we used to come. He would put the child under the donkey. Three times. Because of the whooping cough. Because we were all worried about it at that time. When I was a child, whenever they come home from the war they had TB and the coughs, it was terrible. It was terrible on the lungs the whooping cough. You don't hear of the whooping cough now. You used to hold the children's wrists, you see, and make them cough, you see. But he put them through the donkey three times and he come down for three days, and three days you went up to the Gasworks and the kids were better with it.[25]

We had what we called the kitchen, you would call it the living room, you understand, and there was a range in our kitchen. My grandad was a blacksmith and he used to do us all the different wee things to put on our range; you could set all the kettles and teapot on it and all. And yet the bloody thing wouldn't

May's Market, Oxford Street, c. 1930s.

have cooked anything, we never could get it to cook or never could get hot water. It never heated properly but my mammy could cook on it, and we were able to put our porridge at night in the steamer and it would have been lovely done by the time we needed it next morning. And you always set your table at night, the table was in the middle of the floor, a sofa and a table in the middle of the floor with the chairs. A big cabinet with all the dishes and knives and everything in it. My mammy was very fussy. Tablecloths. You went out to eight o'clock Mass in the morning round to St Malachy's seven days a week.

You had your days for soup and your days for stew, different days you done different dinners. Always a fried dinner on a Saturday. You had fish Friday. The fish was gorgeous. And the herrings. The men would

Rose Ann getting fish from Hughie Smyth's mobile fish shop in Joy Street in the 1940s. Rose is watched by her sister-in-law Lily Smyth.

have come down Oxford Street with a cart, flat cart, a handcart you called it and they used to have a couple of boxes of herrings, and the herrings would have been jumping alive, they were only off the boats, you see, down at the docks. And he would have shouted 'Ardglass herrings'. Everybody was out. Herrings, and you had herrings that day for your dinner and new potatoes. The potatoes were gorgeous. The food was lovely. I don't know what they do with it, that business the fridge and all, it takes all the good out of the food. You know what I mean, anything straight from the ground it was marvellous.

I would eat eight kemp apples a day, you just eat them. See the way they eat sweets and all now, the sweets we ate were the black-brown homemade toffee. With brown sugar. And the men used to come round with that too, come round the doors. And every woman if she had apples and candy she would have done it and then opened her window, pulled back the thing and put on the window to sell them. Toffee apples and maybe coconut on it. Dip them in coconut.

Come up the stairs. We had a back room that was a wee small room, and my Uncle James, he come from America after having his leg off and all, he was in the UVF Hospital here in Belfast. He slept in the back room with one of his sons. Before my Uncle James come, my brothers Billy and John used to sleep in that room and I slept in the return room and my mammy in the front room. None of us could sleep with her because she had TB, you see. She was dying with TB. Whenever my daddy was bad, I never saw him. I only looked in at him at night, we couldn't go in and hug daddy. He was in the First World War and my daddy died when I was going five and my mammy died ten years after him in 1934. I was born in 1918. He had TB, all them men had it when they came back. I never knew my daddy walking.

Mammy was a devil for telling you to take iron, you see. On account of TB she was always telling us and driving it into us what to do. How to walk, how to talk and how to dress. She slept in the front room and my mammy could have saw way up the bridge, Albert Bridge and into the markets. And then the people all going to work and the girls and they all walked, they walked. And the figures they had. Everybody was laughing, full of life, you know what I mean? A different time entirely. And the men used to come after seeing their girls and they would have been over the bridge and they would have been whistling away like a good one, all the love songs and all, whatever you had seen in the pictures and that. It was a different atmosphere and you could have told nearly the time by the attitude of the surrounds. Do you understand? The Markets.

Well then you had the fish market, the wholesale fish market, and then after that you had the cattle pens where they brought the cattle, with the cows going round. The country men, maybe selling their cattle to the butchers, do you understand? Then their docket was stuck on that, the ones they bought. The cattle drovers, cattle wallopers, as I call them, they took these down the Oxford Street. The traffic all stood, ten mile an hour going round Cromac Square. And they would have come out into Annette Street, through Annette Street and into the hay market. As they were going, some of them would have stopped, there was a watering trough for the cattle. Davey Moore, he used to have grass, he sold grass, bundles of grass.

A cow come up our hall one time. My Aunt Dolly, Uncle James's wife, they were only home from America and they had come in to visit, staying with us, and one of them come up the hall! Think of a cow coming up a hall. Wrecked the hall stand and the curtain, and I was coming down the stairs at the time and there's the cow face up with the two horns up. Then a man come and he had to get over, climb over the cow, and talked to the old cow, right. And catch it by the horns and another man by the tail, and that's how they directed the cow out. They were pushing, 'Right, right, boy, right, calm, all right fella'. Out it went. The man gave us money for destroying the hall stand. The best of it was I was telling you about sliding in the hall. You see, the poor old cow slid up, you see, on the mats.

There was a fair day once every Wednesday in the month, I don't know whether it was the first Wednesday in the month or not. It was great, the fair day. Just outside our door, the fair, you see. It was mostly gypsies and that, you know. Annette Street. They all had hats on them, a hat or a cap or something, and they would have walked up and down the place, they would have run with the horse, you see, and the men would have been looking at it back and down and if they sold it, spit on your hand. The watering trough was over there in the street and there was a phone box. No phones in the house, you see, but there was a phone in the '30s in Cromac Square.

The variety market was in front of the courts. The variety market was a variety of everything. You got everything in it. Second-handed clothes, pictures, jewellery, everything. They sold everything in the variety market. It was on a Friday, just a Friday. You paid for your table, and a wee roof on top of it. You paid for your stall. Then if you were against the wall it was great because you had the wall to show your things off and you didn't get half your things knocked off. Now do you understand? Over at the wall because you could watch your stuff.

See your house? Honest to God, the stuff that came round. You never had nothing dirty in the house because the ragman came. The food man come after the dinner, keep that for the scraps man, you see, they would have come for to feed the pigs. In different yards people would have had two or three pigs. Do you understand me? Fresh herrings, they would have come, a cart. Then the baker would have come round, the postman, the milkman. Market Street, the nicest people you ever had was Market Street. It was a family. Everybody loved everybody and if their children come with a dirty nose they would have taken up a cloth and 'What's the matter with you?' You know, looked after everybody's child. 'Cassy, I'm going out, I'll not lock the door.' Nobody locked up the house. The door left on a till and keep an eye there. They would have played housey, bingo. The women would have set the thing, the kids down for the slate and playing bingo. Played on the street, you just sat on the crib on the footpath, the kerb, and hardboards, you had the numbers on the boards, you see, and you marked it with chalk. A penny, a penny or something. Whatever it was you got the pennies. Marvellous, honest to God. And the ice cream when it come round. Angelina would have come round with her ice cream. She was Angelina Morelli. Sold ice cream. She sat on the thing. A penny slider, oh, a big penny slider, a poke. She had the cart, like it was a proper wee cart. She pushed that from down at the docks, down beside my grandad's forge. My grandad had a forge, you see, and Angelina used to live down by there.

My grandad was, oh, a big man and he used to throw the water, the cold water up round him, you know, and he went into the bar at six o'clock – no, ten to six, and he went into the bar, in Campbell's. Campbell's faced onto the markets, in East Bridge Street. He used to have it, 'Right, Mr Austin' and my granda used to have a pint and a chaser, a whiskey or something and he didn't pay for it, he paid it at the end of the week, and that was for Mr Austin. And my granda would have been coming out when he drunk that and put his hand on the door every night at six o'clock, turned the handle of the door and came in, and she was supposed to be coming out of the working kitchen with his dinner to set on the table. And he used to wink at me, you see, do that there, and I used to look. 'Ach da', she used to say, 'Ach da'. 'Get the damned children out of here,' he would say. And my granny used to be sitting by the piano. And she would have been sitting. She was a wee thing, you would have hardly saw her and she used to sit with her feet on a cushion thing and she used to sit with that. Her toes. Never saw her legs or feet or nothing. Oh, you never saw none of her body at all. The skirts would have been round and her wee shawl, her wee totey wee knit shawl and the wee thing on the top of her head, and she would have been crocheting or darning maybe their socks and she would have been singing, she would have started to sing:

> 'Treat my daughter kindly
> Say to her no harm
> And when I die I leave to you
> This house and little farm.'

It was like a city inside a city, you know
Billy Shannon

Fifteen, down to the Ormeau Road, cross the Ormeau Road into the Gasworks. Hadn't been on the Ormeau Road much, someone told me they were starting men down there. So, I have just left school and I remember crossing in through the Gasworks gate and a wee man stopped me at the gate. 'Where are you going?' Says I, 'I am down here to see if I can get a job.' He says to me, 'Have you arranged an interview?' Says I, 'I have.'

So, he was sitting at a big brazier fire which sat outside and he says, 'Come on, I will show you where the offices are'. So, away I went into the offices, this wee man took me up the stairs into this hallway which I thought was unbelievable, all tiled hallway, all green tiles. I thought it was brilliant. In through the door, a man very well dressed, the suit, sitting behind the desk. 'Who are you?', 'What are you?', 'What age are you?' Right – so I told him my age, fifteen, and he says to me, 'Do you know somebody who works here?' which I did, and I told him the fella's name, and he asked a few questions and he says to me, 'When could you start?' 'I suppose I could start anytime.' He says to me, 'Well, tell your ma to give you a piece and send you down here on Monday morning. You have to be in at half seven to eight o'clock and you will punch the card. I will leave word down at the gate to get your name put on the clock card.' So that was my first experience of going into the Gasworks.[26]

They were just bringing new cookers in, it was called the 291 Cooker, a grey cooker, and everybody thought these were great. Wee round discs at the bottom of the cooker which you had to put on with a screw and for the next three months I put these discs on. There were thousands of them. So you started at eight o'clock and by half four I am telling you, you were fed up looking at the discs! That would have been around '60-'61 and Belvoir Estate was just being built then and they got these

Billy Shannon at the Gasworks just after it had closed, late 1980s.

21

new appliances in, a gas fridge, which at the time everybody thought were the last thing. So, you got a job on Monday morning and you would collect your meter, a gas meter, with a piece of lead pipe on the top of it and the van left the cooker up with the rest of the appliances to the house. Now the van was only a new thing, but you got on the trolleybus, up to the top of the Ormeau Road, off the trolleybus with your mate, who was the gas fitter. I was an apprentice gas fitter – 'the Get' – in other words, for to get him this and get him that! So, you drew the fittings out of the store, he carried the meter, I carried the tools, off the bus at the top of the Ormeau Road – that's where the bus sort of turned, you could go no further – and you walked from there to Belvoir Estate, whatever address it was and you started.

It used to be all lead pipe, but this was the first time they were doing everything in copper pipe and everybody thought this was great, this was the new thing, copper pipe. Lead pipe was beginning to go out. By the time you got the cooker in, the gas meter in, the gas fridge and the Main Minor in – it was a small gas water heater which gave you instant hot water – that job usually lasted a full day. If they wanted a gas fire fitted, we fitted the gas fire as well. Everybody thought this was great at the time. You had to keep your bus tickets; when you came back that night you filled a wee docket in to say that's how much it cost you for to get up there and get back and you claimed your money back. I must have worked up at Belvoir for about three years. It's quite a big estate and when we first went up, they were only beginning to put the roads in. All brand new houses. From living in a wee house in the Shankill coming up to this. You were getting quite a fancy house, a nice wee driveway into it. Brilliant, you know. I thought it was brilliant.

All the appliances were from the Gas Department showrooms, there was one at the bottom of the Shankill and there was one on Ormeau Avenue. People went in and picked their appliances. You hadn't an awful lot of choice then. Gas fridges, gas fires. What other gas appliances? Mantles, gas mantles for light were beginning to go out, but you were still doing gas mantles in houses. We still had the parts and all down there. Gas irons – at the side of the cooker you had a wee bull nozzle, and what you did, you pushed your hose up onto that and you turned it on there, and the tube went to the iron and that would have given you the heat to iron clothes. Again, they were beginning to go. Gas pokers, you used to light a coal fire with them – at the side of your fireplace a wee nozzle and a red-coloured hose to a long thing, about 12 inches, and on the top of it, it had rings, spiral rings going down it, and jets, just open jets, and when you actually lit that there the flames came out and you pushed that into your coal fire to light it.

You had places like the Ormeau Bakery. Now there were all gas meters in there and when you think of a gas meter, a normal gas meter was about 14 inches by 14 inches for a house, but you are talking about a gas meter maybe five foot high and five foot wide. They were one of the big users of gas. The crematorium, it used to use the gas. Most of the big firms were supplied with gas. The shipyard was the same. But gas street lighting was just starting to go. It was unbelievable when they were taking the gas street lamps out – they were cast iron and quite heavy – they used to take them away down to the back of the Gasworks to the scrapyard and there were a couple of men used to break them up with sledge hammers. And then the scrap man would have come in and taken those away. It was hard to believe how much scrap was actually down there.

The Gasworks was very, very big. It was like a city inside a city, you know. Barges used to come round from the docks, bringing coal. They came up the Lagan and up a wee river at the back of the Gasworks, towards Cromac Street. The barges used to come up there and they used to pile the coal in mountains using a grab which went down into the barge. Then there was men standing there with shovels and they would have shovelled the stuff over all, into barrows, barrows with four wheels, and would have pushed those away to throw into the retorts. When I first came in they were still running the retorts, they made gas down there. The coal was thrown onto conveyor belts that brought it up and into the retorts. The retorts were massive big buildings where they heated the coal, extracted the gas. And then they were left with tar and cinders that was all threw out the other side and the men again had to throw that into big barrows and into another mountain. Then there was the big pit down there where the extract of the tar all went into. When I first started, people used to bring their youngsters down who had whooping cough. If they put their head over they could get the fumes coming up which was supposed to help people with whooping cough. A big round tank of tar, it was hot and the smell that came out of it! It was hot coming out of the furnaces, it was steaming up like a bath – no health and safety then!

But it was the end of the era, they stopped making gas down there and it was all piped in.[27] When I started there was maybe five to six hundred men working there. There was not much mechanisation then, but just when I started in the '60s everything was beginning to change. When I started it used to be seven days a week. Then, like everything else, they tried to save money by mechanising it: they cut the number of men down, cut the hours down, though there was always a night shift and a day shift. It used to be when the people were feeding the retorts, they used to be allowed to go out for a pint every couple of hours because the heat in those places – it is hard to believe the heat in those places. What they used to do, they used to go out through the wee yard, punched their card and were allowed to go out to the bar across the road. Twenty minutes you had to go, get your pint, come back and start back at your work. Not much health and safety there! It was a thing they were allowed to do because I think it was so warm where they were working. I am talking about men with sweat running down their faces. Black as your boot sort of thing. Really, really black. But again, mechanisation was coming in there, everything was beginning to go to money. Trying to cut down on things, you know what I mean. They had horses and carts to pull stuff around the yard. They had actually stopped using them for outside, they just used them for pulling any big pipes. You see some of those pipes coming into the holders, you are talking about maybe 24 inch. Massive big pipes like, you know. They used them for that or for carrying anything round the yard. They had stables down there and they used to have a couple of men then who looked after the horses and the stables. Again, that was going out when I started.

Great atmosphere about the place, like a wee town, you know. They were from all parts, but most of the people were from the Ormeau Road, I would say. Donegall Pass, I would say most of the people who lived in Donegall Pass actually worked at one time in the Gasworks. It was all men, no women. The only place you might have got a woman would be in one of the showrooms selling some of the appliances. Just beside the clock tower, as you went in the main gate, just over to your left, you went up into the offices. That is the building that is still there now. That was all the managers, all the leaders was up there, all the workers were sort of way downstairs. When you went straight on into the Gasworks, you came round to the retorts and then the holders. Then down past the holders you had the electric shop, the fitters' shop and down at the very bottom you had the blacksmith and where you stabled the horses. That was coming right down towards the Lagan. In the Gasworks was 30 acres. There was arc lights all round it but it was only in the '70s when things began to turn here and you were beginning to get a lot of trouble that they really lit it up at night. The army was stationed in it then. I think there were six gas holders and one massive one, the big grey one. It was built later. The difference between it and the other holders was that the outside of it didn't go down, the inside went down. So where you had the other ones the whole tank went down, but in that big one it had like a plunger inside of it. There was a lift right up the side to do the maintenance on it. No matter where you were in Belfast, you could have seen the big grey gas holder. But as I say, like a town.

There were cats down there, thousands of cats. When you went in through the door if you had looked over to the other side you would have seen all the cats being fed at lunch time. The men threw bits of their pieces down. And I am talking about hundreds of cats because there were hundreds of rats in it. You had the rats coming up off the river and someone at one time had brought cats in and it's like everything else: they must have bred, bred and bred. Some of them were really feral cats, wild.

A hooter, a big, big blower went off in the morning every day at eight o'clock. They told you to be in before the hooter went off, but as long as you had punched your card before eight o'clock you were okay. A big machine with a clock on it, you pushed your card into the top of it, it registered the time. You were allowed five minutes after eight but after that five minutes they took the first half hour's pay off you. My pay in the Gasworks when I first started was three pounds ten shillings for a week. And I remember bringing that home to my mother, and she gave me the ten shillings, and I thought I was on top of the world!

Why did it shut? Well, they said it wasn't paying them for to make the gas there. Everything started to run down. They started to go to electric. Everything went to electric. There was the idea of bringing gas up from the South of Ireland, but they didn't like that, a lot of people didn't like that. It was an option at one time. This was natural gas, they had it piped in to Kinsale and they were talking about bringing it up here, but then that didn't suit a lot of people because they didn't want the green gas coming up. I remember

Right: John Kelly barges delivering coal to the Gasworks, 1964.

The Gasworks and Blackstaff River, 1987.

they had the meetings, you know, the shop stewards and all was there, 'What about bringing this up?' Everybody who was in the Gasworks, well, a lot of people, wanted it up. But then there was a lot of other people who didn't want it up, you know. Politicians didn't want it up. And that was the death knell then.[28]

I was there until 1990. About three years previous to that everything had started to run down. They had stopped bringing appliances in because they were on the rundown. And I actually got a job in the stores doing bits of this and bits of that. At the very end there were maybe 30 or 40 people down in it. Well, you take the town, it's not a big town, this, but you take all those gas cookers, all those gas meters. They all had to be all lifted from the houses, you know. Every house had to give their gas cooker up, gas meter up, and they capped the gas at the service coming in. All that had to be done which was a big job, like, you know what I mean? It wasn't done over, say, six months, it was maybe done over five or six years. Because when you take all those places that had to be converted, like Ormeau Bakery, those meters all had to be taken out, they had to give them notification because they had to convert to oil or bottled gas. Bringing all those appliances out of houses. Again they were brought in, everything scrapped. All threw down at the back of the Gasworks where previously the old gas street lamps were broken up. Think of all the lead, scrap lead, that was coming in there too. All that had to be scrapped. A big, big job, like, you know.

I just got very interested in the markets when I was a kid
Paddy Lynn

My mother was originally from the Markets and my father was from the
Half Bap which was behind St Anne's Cathedral. And they bought a
house in Donegall Pass and they had a family. We moved out of Donegall
Pass at the start of the Troubles, around about 1970. We moved back to
the Markets where actually my grandparents were from. My great-
grandparents were from that area as well. So I grew up in the 1970s
around the Markets area. As a kid my playground was the Markets. I
remember I used to work every day after school and all day Saturday in
the cattle market. Washing out the pens. After school and before we
went to school, we used to go over and clean out the pens and help the
farmers bring the livestock. It was where the Waterfront Hall is now.
Colgan's and Allam's cattle mart was in that area, by the Belfast fish
market, which was the sister market to St George's. So I worked around
there. May's Market was massive. There was markets in it two days a
week. That whole complex is now the Waterfront Hall, the BT Tower, the
Fujitsu building. That was all May's Market. In behind it was May's
Meadows. I grew up around that part of the city.

Paddy Lynn, 2011.

But I did not just work in the cattle markets, I also worked in the fruit stores, Maggie Kelly's fruit store.
They were wholesale and had a stall as well. Farmers from outside the city would have brought
vegetables, potatoes and so on. Then you had Donnelly Brothers who specifically dealt in potatoes. Every
week there would have been a market on and a lot of local produce would have been sold. Maggie would
have asked us to help her. But we would have rented a stall and got a loan of produce and sold it at the
market, paid her once we got the money and shared the profits out of it between two or three of us.
So there was a number of things that we did.

It was a great place to grow up. There was a lot of interesting characters, a lot of interesting stories. Fond
memories even though the '70s were very difficult in Belfast. That was my playground, Oxford Street and
all the markets, there was ten markets in that area. The Belfast variety market, Belfast second-hand
market in Lower May Street, they were actually right next door to one another. The old ramshackle stalls
that were made like a lean-to with felt roofs on them. They were part of our life because my mother
worked there. That was where we got our clothes, even for the summertime. My mother worked there
every Friday. She used to work in Donegall Pass, in Emerson's second-hand clothes yard. She used to
buy clothes off Mr Emerson and bring them to the markets. Smooth and iron them and bring them to the

market. Put them on rails. So she grew up with all the other second-hand traders: Maggie Gault, Sally Elliman, Mickey Anderson. Although they were much older than me, they all had stalls, I used to help them out, give them a hand.

In 1978, they moved all the markets across to St George's. They merged all the different markets as they declined. They merged everything: the fish market, poultry market, clothes market – the second-hand clothes market, variety market. So it was all merged into one, into St George's. Fifteenth of November 1978, that was when it moved over.[29]

The rag-and-bone men bringing in all the clothes, before they brought them for recycling they brought them into the market and people got the pick out of them. Before charity shops and all, people were going to the market to buy clothes. You know, you could have picked anything up. 'Oh, my son's going back to school in August, we will go down and we will buy some shirts and stuff.' Most was near new because people took care of clothes. It was largely a working-class area, so largely working-class people, though you know all classes came to St George's. Middle class, upper class and working class. And those were the kinds of things I enjoyed because of those different types of groups of people. And that was unique for Belfast as well.

Allam's cattle market closed, Colgan's closed shortly before that. So slowly Belfast markets were in decline. They were starting to slow down. It was a massive industry at one time, market trading. There was thousands of people employed in it directly or indirectly, supplying stuff for people. Belfast is a completely different place now to what it was then, the place I grew up in. The Troubles had a massive impact. Lots of people moved away from Belfast completely. It was inner city, the markets were inner city, you were three-minutes' walk from Ann Street, you know what I mean?

The redevelopment of the Markets area started in 1975/76. I came from East Street, we moved next door to my granny in 18 East Street, facing the old hide and skin stores where the hides coming off all the cattle that were slaughtered in the local slaughterhouses were brought and they were cured so they could be used. My family didn't move out of East Street until 1980. That community atmosphere that existed back then was brilliant. Then the Troubles sort of destroyed a lot of that as well. A lot of people had to be a lot more cautious, you know. But the Markets was very mixed as well. There was people from many countries of the world lived in the Markets. Although we got a reputation because of the Troubles of being quite a nationalist area, we had neighbours who were Protestants and we had neighbours from other parts of the world.

What took me into the markets? I don't know. I just got very interested in the markets when I was a kid, I just got a buzz out of it. You have to be a people person to be a market trader, and I think I learned very quickly that it was important to communicate with people and enjoy life. When I first got my own stall, we

St George's Market, 2011.

used to do it every day after school, so it was something that just stuck with me. There were loads of young people. It was a way of making money, you know what I mean? My mother had a stall in the second-hand market, the flea market. It was by the Law Courts, next to the variety market. I first helped my mother then I got a stall of my own in St George's doing second-hand bric-a-brac. But that wasn't until the mid-1980s, you know.

I first got involved in 1990 with the National Market Traders Federation. I was involved in community groups in the Markets, I was a youth worker in the Markets in the late 1970s, things I enjoyed doing, and I got involved in the National Market Traders Federation because the council had proposed to move the traders at St George's to the car park at Smithfield and we ran quite a successful campaign, over 40,000 signatures. They wanted to turn St George's into an arts and crafts centre but it was the last of ten markets. I thought this was very unfair and so did many other people – an area without a market, yet called the Markets community because there were ten markets there at one stage. It would have been very sad for Belfast. So we went along to the council and we said, 'Look, we don't agree with you!' We went along and said, 'Why don't you do some work to it, remove the old roof, do the place up, and leave the market where it is. And we will support you in an application for a grant.' So the council applied to the Heritage Lottery Fund with our support.

But it was a hard struggle because at one time the councillors were not interested. And what was worse, most of the councillors that actually wanted to move the traders from St George's never came into the market. We found that very strange. And we ran quite a successful campaign. And ach, well, it just grew from there. We worked very hard. We meet up weekly with council staff. We have sorted all the problems out – well, there are always going to be problems. Like, you must remember there are 20,000 people a fortnight go through St George's Market. Big numbers coming through the doors on a yearly basis. It's become a massive tourist attraction. And it's great to see that, beside the Waterfront Hall. The old and the new of Belfast can coexist.

When it opened in 1978, it was on a Tuesday and a Friday. Now it's Friday, Saturday and Sunday. We extended, we have the Sunday permanent now, the council approved it. We hope at some stage – although our population is very small – to sustain a seven-day-a-week market. Hopefully sometime in the future, St George's can be open five, six, seven days a week. But it's great to see the market being used for other events as well. Fashion shows, boxing tournaments, motorcycle shows. You name it, St George's has been used for it. It's great it's not just used as a market, it's a meeting place for everybody and it's a place for all uses.

In Murdock's yard in Verner Street they still have some horses there. Trotting ponies, the stables are in Verner Street, the lane just at the side of St George's. They go to different competitions. And they have been asked over the years to sell their wee place and they haven't done that. And I am quite pleased that they haven't done that. There were so many of those wee yards where there were horses and different types of livestock. That's the way Belfast was at one time, the inner city of Belfast. But the horses in Murdock's yard, they must be the last horses in the inner city of Belfast. No doubt about that.

And then you had Barney Ross's yard, it was in another part of Verner Street. He had a rag yard, recycled clothes. And he used to hire horses out to the rag-and-bone men. And the wee rag-and-bone men, the like of Davey Campbell and old Paddy McFarland and Mick Anderson, many of those old fellas that I knew, they rented horses and carts out of Barney Ross's rag yard. They loved those animals very, very much. He would rent you a horse and cart. You went in in the morning time, but you had to show him that you could control the horse and you were responsible for it. Rags were the thing, and he would have bought the whole lot off you. When I was a kid, he had about five or six horses. But he only rented them to people he liked. It stopped in, oh, the late 1970s. I remember that well. I remember going round and getting bits for my bikes and stuff, he would have taken scrap metal and lead, copper, brass, he took many a thing. But I grew up round there, bringing bits and bobs in. Always out trying to make a living.

We got the old lock-keeper's house
Alice and Jackie Murray

McConnell Weir and lock, 1985.

Jackie. What happened was when there was heavy rain, they would have got in touch with the police and the police would have come out and said you were wanted down at Queen's Bridge pumping station. The police would have called because not everyone had telephones. My job was to check on the pumping station, there at Queen's Bridge. There was another pump house on the Lagan here at River

Terrace, but there was a man working all week at River Terrace and was on standby all the time. I had no car, and if it was teeming out of the heavens you still had to get to work. I would have been on standby, if the pump tripped out for to reset the pump, you know. Pumping the water into the Lagan. Then the man at River Terrace retired and they handed me the job and we got the old lock-keeper's house at McConnell Weir here in Ormeau.

Before we had this house, it was where the lock-keeper lived. I heard that the lock-keeper would have been walking up and down here in all his regalia like a petty officer, you know. Walking up and down as if he owned the joint! He finished work 41 years ago, when we took the house on. He operated from the house here, he had one room as an office, the room that's now our dining room had like a big wardrobe, casing with all the electrical equipment. In the bay window there, he had a big working table. He was away before we came, but there was an electrician who had to do with the maintenance of the locks, he would have came every month or something and went round all the electrical points and would have greased the locks. There was a weir right across the river, the McConnell Weir, and by our house a lock to let boats go up and down. I remember the coal barges, one coal barge with an engine in it towing three, four barges. When we moved here the locks were there but they weren't in use.

Alice. A couple of people had houseboats on the river here. James Young, the comedian, had a friend who had a boat here. James Young would have come down on a boat, and this friend was an actor – what was his name? A tall dark man, he was an actor too. And there was another man called Mr Sprote, he had a boat tied up at the lock. His wife taught in Stranmillis College and I think he worked in the aircraft factory. They used to come for their mail here, their mail was delivered to our house. Sprote had a wee young boy and a tall boy. Very nice people, I must say. One night when the Troubles were very bad here, I think 1971, he said if it got any worse, he would take us away on the boat! But then we would have got mail for them from different places and we had a few gentlemen calling to the house because – wait for it – all the big shops down town would have been looking their money but the bird had flown! He had gone! Delicatessens, food bills and all started coming!

Jackie. The lock just went into decay, and then when they built the new weir, down near Queen's Bridge, they got rid of the McConnell Weir. They told us to go away for the day and if there was any damage done when they blew it up, not to be alarmed. But it just went up with a wee spit not a big bang. That was the wall broke, the weir broke, once that happened they cleared it up. They kept the lock and part of the weir is still in the river.

Alice. The lock house is a sizable house but it's a bit of an illusion, so it is. It's a three-bedroom house with a garden, just a family house. There were bungalows when we first came here, prefabs were here when we first came, we were kind of hemmed in. A real community then, but as people moved out they got

Right: The McConnell Weir, Gasworks and the white-painted former lock-keeper's house.

houses, better houses elsewhere. I don't remember anyone from the bungalows staying, everyone moved out. The Gasworks was still running when we came here, we were here the night there was a bomb and it blew up. Horrendous, wasn't it? Absolutely horrendous.

Jackie. The tank getting redder and redder. We were stood watching it, people were out from their houses watching it. So then it got that red we decided to move down, you know. We got halfway down the street when, just bang! We felt the heat on our backs. I had a car sitting parked out here but never thought of the car, just running for to get away from this gas tank getting redder and redder. Eventually a bang and you would have thought it would have blew down half the street, you know.

Alice. It was horrendous and there were loads of people here that were invalids and all. I don't know how there wasn't anybody fatally injured here. Horrendous, wasn't it? Really, really frightening, terrible. It was a bomb and there wasn't the usual amount of gas in – they said it could have knocked all of Belfast down, I did hear that. There was only the dregs of the gas that was in it, and believe you me, it was horrendous. We were lucky that night, as we had two or three of the children over at my mammy's. The whole far side of the street, we felt it running up the street, a swish of fire, you know where you see a volcano, swish, right down the whole street.

Jackie. See, we had gone down the street for to watch it, it was getting redder and redder.

Alice. The army came in and told us all to get out. I'll never forget it until God calls me, you know. Flames came right the whole way down the far side of that street, you know. And when you turned to run you could actually feel the heat on your back. It was just like a flash and that was it but it's still a miracle that nobody was killed by the flames. There was actually no help here, nowhere for the people to go, but a church opened up a hall and got people into it. We went over to my mammy's, stayed over at my mammy's and came back again, you know. The house was all right. I still can't believe it.

Now here's a story: two little boys were playing one day out here by the lock and the next thing we hear is, 'Wee boy in the Lagan'. Now Jackie couldn't swim and I couldn't swim. Not a life belt out there. My son wanted to jump in. Now, I know he could swim but he was only about 14 at the time and these kids were about eight year old. One had a frog, and the other one says to him, 'Let us see it'. He gave him the frog and the frog jumped into the water and the boy pushed him in to get it! That's true! So, I ran out in the street and across the road, 'Somebody's in the Lagan! Somebody's in the Lagan!' Jackie went out into the yard, got a brush, a yard brush, put it down here and he was able to pull him out. He got him to hold on to the brush and he pulled him out with the brush. He got out and all. Jackie went to use the brush later on, to brush something, and the head came off the top! So, like, it was touch-and-go.

We did the oatcake
Lynda Young

I went into the bakery in 1968. I packed biscuits, put them on boards and you had to weigh them. There was all these different biscuits. Digestive, lemon, signature, different ones. The signature ones were nice, a round biscuit, not sweet. They had a butterfly on them, the shape of a butterfly. It was the crest of the Ormeau Bakery – it was on the vans of the Ormeau Bakery. We did the oatcake, there was a machine for the oatcake. Round and square. When Marks & Spencer came over they got an order and we did a lot for England and exported to Canada and Australia. Sometimes to Europe.[30]

I worked in the biscuit department. There was the chocolate room, toffee room, cake department, the dairy and a place – what was it called – where they did the wedding cakes. Breadmaking was downstairs. They did chocolates, they were gorgeous. Wee boxes, but very sweet. The dairy brought in milk for the bakery. They owned their own farms. We went one time and we looked round all the cows and all and then afterwards we had our tea. The farm was somewhere at Lisburn. We just went one time, we just went for the run, really.

In the biscuit room we made and packed biscuits, then about a Monday you did chocolate digestive and the chocolate snap. Well, you put the biscuits on the machine and it coated it and it went through a cooler and came out the other end. The biscuits were put in packets. We went in the side entrance, never the front entrance, and you had a shower first and clocked in. There were cubicles, about eight in our floor and about six in another. Then you changed into white clothes. There was a Miss Johnston who changed the towels on a Monday, and if your towel wasn't wet on a Monday, well, you had it so. You always had it wet. We had white clothes with a blue collar. In fresh cream it was green, for chocolate it was brown. White clothes with different collars. Your hat, oh, you had to have your hat, but it was very warm. Then they got net ones. Thirty-six worked when I started, then, unfortunately, it progressed down to about 18 then about 11 when I left. With machinery, they got different machinery and didn't need the people.

When you went in the main entrance it was lovely. You thought you were somebody. There was a big staircase. For the public, you had visitors come round to see round Ormeau. What was it like? Oh, you know Stormont, the big stairs there? It was lovely. Made of stone, marble. There was a shop at the front but it was not connected with the bakery. It had a swinging door on it, it was lovely. They had about 14 shops, there was one up at Newtownbreda and there was Fountain Street, different ones. They had wee vans that would deliver to your house. They used to be the electric ones, then just the ordinary ones.

The Wilsons owned Ormeau Bakery, they lived up the Malone Road or Lisburn Road. They came round periodically in the mornings. And they were, you know, very stately. A chauffeur-driven car. They would lift

Ormeau Bakery was sold in 2002 and the building has been developed as apartments.

the biscuits to see whether they were all right. There was a shop it was called the half-price shop – you could go down and get whatever was left. It was for people who worked in Ormeau. We thought it was better than any other bakery. Their apple creams were beautiful, pastry. The chocolates. They did sweets, toffee and fudge but the fudge was very sweet. Croissants. They used to do export to different places.

There were flour lofts on the top floor – they made flour, they made ordinary flour, self-raising and wheaten. The flour came in a big lorry, they put it in with a big hose. It was sold in bags and used in the bakery. Below that was the hot plate where they did soda bread, potato bread, pancakes. Very, very warm. It was automatic; halfway the machine turned it automatically, and when they came out the other end it packed it. Then we were the biscuit department and then downstairs was the bakery. We were with the chocolate room and then they did wedding cakes, beautiful cakes, really lovely. Christmas cakes. Sandy Wilton and Dessie Eccles were the cake decorators. The cakes were very expensive. I remember one time, it was an order – they used to do export to different places – and I remember them doing an order for a daughter who went to Roedean School, and they had to have cakes and it had to be, you know, so-so.

I worked 40 hours, five days. But then when you started to work overtime and that and do double time on a Sunday, it wasn't worth your while. You got a break, there was a canteen. But you had to have a ticket, you weren't allowed to have money. You had to go down and get tickets. You weren't allowed money in your possession in case it got into the food. I got a shift, I started at five o'clock in the morning, but I stopped at two o'clock. But some of them started at eight, some of them started at nine, it just depended. It was nearly seven days a week, that's why I just couldn't do it, the shifts changed and I was coming in at nine o'clock in the mornings sometimes and going away at half ten at night. I couldn't do it, that's why I had to leave, unfortunately.

All the swings in parks were chained on Sundays
Alan Wilson

I suppose the best place to start would be when I left school. I left school when I was 14 years of age. Our school had a cloakroom, and I remember announcing to everyone that I was not going back again. The brother and I both went to the same school, he stayed on and went to Queen's and then eventually into teaching, but I was not terribly interested in staying, so I left. Whenever I left I didn't know what I wanted to do, but I got an interview for a job in the shipyard. I had this grand idea I eventually wanted to be a trainee draftsman and I went for an interview and I remember it as clear as anything. I was interviewed in a room in the engine works, and for some obscure reason, the teacher had written on my last school report that I was good at algebra, and the chap who was interviewing me set me down this problem to solve, and to my amazement, I managed to get it right! But anyway, to cut a long story short, he told me that 14 was too young to start and that they would send for me whenever I was 16.

Alan Wilson at the Tropical Ravine in the mid-1960s.

So, at that time, my father was in the parks, he was 45 years in the parks, and I had also two uncles with long service, one 42 and I think the other about 41 years' service. They were all park rangers. So my

37

father spoke to Archie Graham, he was the assistant director and curator of the Botanic Gardens at that time. They got me an application form and I filled it in and eventually I went for an interview and I can remember the day as clear as anything: it was a Friday in the Botanic Gardens, and my father went over with me and George Horscroft, he was the director of parks, he interviewed me and told me yes, I could start on the Monday, and he actually brought me down to the bungalow in the Botanic Gardens and introduced me to the staff in Botanic. That was 1953.[31]

At that time I wouldn't say I was particularly interested in gardening, if I am honest about it, I just took it as a job. I was an apprentice gardener. I liked the men, they were all very decent and I have a lot of happy memories. My first job would have been making the tea, which I didn't really mind. I would go up for the milk at Bamford's Dairy, in Colenso Parade, work in the greenhouses, take messages to the City Hall. The wages were in actual fact not too bad in those days. I was pretty fortunate, I was getting two pounds ten shillings per week, which was reasonably good in those days. You worked a 42-hour week, which included Saturday morning. Most of the men working in Botanic had come from old estates round Ireland, there was a mixture of all sorts.

So when I became 16 I did get a letter from the shipyard saying would I like to start and, after a lot of soul-searching, I decided I would stay in parks, and I have worked here all my life. I remember one day in Botanic Gardens, Archie Graham, who was a bit aloof but a very nice person, he gave me a syllabus for Greenmount Agricultural College and he asked me if I would be interested in going. So I remember taking it home and showing it to my brother, and he said go, of course you will go, and I did. I must admit, it was one the best things I have ever done and I enjoyed it. George Horscroft retired, and the next director of parks was Reginald Wesley. He was another Englishman. I got on with him very well, but he could be aloof and possibly a bit of a snob; I am not saying that in a disrespectful way, I must confess I got on with the man very well. He came in the mid-1950s and did not retire until 1971 or 72. Then Craig Wallace was the next director. In those days there was not the same amount of bureaucracy that there is today, and they would have been more hands-on, though they would not have interfered with the actual work. There would have been site visits, and they would have said to you what they wanted. If they saw something wrong, they would have told you there and then. Nowadays, there is a lot of bureaucracy. In those days the parks department was very big, they had their own tree squad, they looked after all the school grounds and all the housing estates in Belfast, as well as the parks. A big department, but not a lot of money in those days.

There was a building with a clock tower at the entrance to Botanic Gardens on the Stranmillis Road. One of my jobs on a Wednesday was to stand up at the clock tower with a key to the wrought iron gate to let in a man from Sharman D Neill, the jeweller's shop in Royal Avenue. He came on his bicycle every Wednesday, climbed up a ladder and through a trapdoor in the building to set the clock for the week.

The Palm House, closed and restored from 1977 to 1983.

The park rangers had a peaked cap, a proper uniform. Navy blue suit. Park Ranger was the name on the cap. They would have had a whistle, they would have blown the whistle to tell people cycling to get off their bike in the park. Whenever I first started, there were two rangers at Botanic, a fellow called Forsythe and a fellow called Snodden, and they looked the part, you know. They earned a bit of respect. One of them actually lived in the gate lodge in Botanic Avenue. Their duties would have been opening and closing gates and patrolling round the park to see that people sort of looked after it. They would have had a stick with a point on it for litter collecting. They would have got to know a lot of people in the park, you know. Looking back on it all, I would say people were more respectful. I know you can look back and think things were better in the old days, but I do think students would have respected the parks more than now. A big thing in Botanic Gardens would have been someone putting a chamber pot in the hand of Kelvin, the statue, once a year for Rag Day. That was a very daring thing! That would have been the most daring thing that would have happened in the park.

Everything was stored at Botanic Gardens, it was the central stores for the parks, and whenever I started I remember clearly that a horse and cart would have come from the City Cemetery on a Friday to collect stores. There were stables in the yard at Ormeau Park and also still a horse used in Ormeau Park. The foreman lived in the house in the middle of Ormeau Park. A fellow called John King lived in it. And then David Montgomery, he moved in there. They didn't seem to mind that, there was never any problem in getting people to live in the parks. And then there was another house, just at the Ormeau Bakery entrance, there was a cottage there. I remember that as clear as anything. Just outside the wall at the bowling green. Ormeau Park hasn't really changed much except down at the soccer pitches. I would say in the '50s and early '60s it was all a bit of a quagmire because during the war, I think there was an army base down there and all the drainage was disrupted. The golf course hasn't changed, no, but there used to be cricket pitches. There was one really good cricket pitch, and cricket was very popular in, I would say, in the late '60s, early '70s, it was by Ravenhill Road. There used to be, I think, a Wednesday evening league and a Saturday league.

Ormeau Park had swings, and they were closed on a Sunday, they were chained up, chained and a lock put on them. Pulled together and chained up onto the frame so no one could use them on a Sunday. All the swings in parks were chained on Sundays, that was just the way things were, you accepted it. The change first took place at Drumglass Park. There was a campaign. Then it went to committee and to council, the argument was that somebody wants to go out and play golf on a Sunday nobody says you can't do it so why shouldn't people be allowed to use swings on a Sunday? The swings in the parks were chained at night too, closing time in parks and things were chained up. That was the way it was done, and then obviously someone said, 'Why, why bother doing that?' Definitely that was done because my father, he was a park ranger, and I know that's what actually happened. It was one of his jobs. At closing time a bell was rung in each park, a bell high up on a pole, you pull on the chain. Quarter of an hour before closing or something like that. The bells are still there in Ormeau and Botanic.

When I started, down the East Walk at Botanic, near Botanic Primary, there was an aviary looked after by staff at Botanic. A green, wooden structure with netting. Another thing I remember from the 1950s was when a big elm tree blew down in Botanic in a severe storm. There was a lot of damage and debris, and two men came out with cross-cuts, one on either side of the long saw. It was amazing to see how they

Temporary pavilions constructed by the Lagan at the bottom of the Botanic Gardens for Ulster 71.

were pushing and pulling on the saw, the sawdust went up in the air, you know, and it was incredible how quick all the remains were removed and, unlike today when you bring in machinery and a lorry, there was no disturbance to the soil.

All the greenhouses and the Palm House were heated by coke. That was one of my jobs. There were two large boilers down in what they called the bottom yard, behind the Palm House and they were below ground level, which meant when you got a lot of rain, invariably, they would have flooded, you know, and one of the jobs was to help to bail it out with buckets. Throw the water out. And the men would have come down in the morning and de-clinkered the boilers – great big clinkers – after they had been stoked up all night. Then they would have thrown the clinkers onto the ground, big, long shovels with a big, long arm on the shovel and threw it up into the yard. I would have wheeled the clinkers away and then eventually someone would have loaded the clinkers onto a lorry and if they were doing drains or anything like that, they would have used them in the backfill. The boilers used coke from the Gasworks, there was another boiler at the bottom of the Tropical Ravine. It was labour-intensive work, stoked up in the morning, stoked up again early afternoon and again before they left. Seven days.

Ulster 71, that was a big thing. It brought a lot of people actually into the park, they did come from all over. What is now Queen's PE Centre down there, they had funfairs and various things, all that end of the park was taken up with stands and various exhibits. It was, I suppose, a bit about the history of Northern Ireland, really. They used to have concerts down there. Looking back on it now, I suppose it wasn't anything great, but it was a big thing then, Ulster 71. I think they were also showing things that had been made in Ulster, inventions and things like that there.[32]

There was very little brought in. I know today everybody is talking about conservation and we are all going green and all that there, but in those days it was conservation. There was very little waste. Maybe you got for the year half a dozen bales of peat. And that was all the peat that would have been brought in. The men in those days would certainly have been very resourceful. The leaves were all gathered, they were put into heaps and they were turned regularly during the winter. Leaf soil. And the potting compost would actually have consisted of a lot of leaf soil. There would have been no peat used at all, really, though nowadays, it's all more talk about it than actual conservation. That's my personal view, you know.

The horticultural displays in Botanic were excellent, quite magnificent. A lot of the plants were grown from seed. A lot of the plants were labelled, even out in the grounds plants were labelled. There were beds laid out with plants arranged scientifically up by the bungalow, the land lost for the Ulster Museum extension. The Tropical Ravine and Palm House were beginning to get into disrepair, but whenever Craig Wallace became director, he seemed to have a lot of influence and was able to get money and pushed to get those buildings restored. It was a very good job. Looking back on those days, I would have to say there is less planting now than there ever used to be. More money but less work done.

I'm river manager, but I still keep my hand in with the diving
Peter Gallagher

The River Lagan here at this particular point is about 150 to 160 metres wide, and you have five hinged, fish-bellied gates, each gate was made in Harland and Wolff when the weir was built back in '93. Each of those gates is bottom hinged and can be moved up and down from minus two metres right up to plus three metres. Plus three metres is actually the quay level around there. So if we have the gates fully up, they are the same level as the quays, and in certain conditions, we can keep out a high tide. But the primary function of the weir and the reason why it was built was because of the old mudflats, the way they were in the past didn't make the area attractive for development or regeneration. So basically, the gates keep a minimum pond of water upstream at all states of the tide. The weir was finished in '93 but maybe signed off in '94.

The Lagan Weir is an interface between salt water and fresh water. They could have made the Lagan Weir like the old McConnell Weir, which you will notice at the back of the Gasworks – it was a fixed-crest weir with a movable gate in the middle and a lock. But basically with the old McConnell Weir, what used to happen was you had salt water meeting fresh, and the salt water, which has a slightly higher relative density, would come in, over the top of the old McConnell Weir, and it would sink gradually to the bottom and form a layer, upstream of the weir, which would be called a saline wedge. Basically, that layer of saline water would stay there forever, and the upper layers of fresh water would go up and down. There would be no agitation, it would stay as a layer on the riverbed, all the organic sediments would suck the oxygen out of that layer and make it uninhabitable for fish life. So that was the main problem with the old fixed-crest weir. When they were developing the new Lagan Weir they decided to try and destratify the water, use the design of the new weir to mix the water a bit better. So you have five movable gates and in addition to that, in each of the piers of the weir, you have two very low-level sluice pipes with a flap valve on each end. So basically, what happens at certain states of the tide, we can, at a very low level, sluice water out and get a bit of circulation going to get rid of the saline effect.

We have the Atlantic seals, they would be up and down quite a bit, we have some cracking photos of them going over the gates. You have guillemots, black-headed gulls, cormorants, redshanks and plenty of others. Oystercatchers you see quite a bit up at McConnell Weir, they would all be sitting in a wee row along the weir. A couple of our boatmen are very into the wildlife, as I am myself, I am trying to learn a bit more about it. We have done a tern island up the river, a wee floating one. We went down to the RSPB and asked them how to go about it and we got cockle shells, and the boys built the raft out of recycled wood and basically floated it down the river. We just put it in before the nesting season last year and just missed it. But I think this year we are certain, you know, to get a few. We have constructed a duck raft as well, which we are putting in opposite the tennis courts in the upper reaches of the river. We tend to get a

Lagan Weir, 2011.

lot of ducks, but they never get above a certain size, a lot of them tend to get killed young, so we thought the raft might help, you know, keep them away from predators. What else do we get? We have a healthy salmon population. I have some cracking photos of a seal eating a salmon down there.[33] There are fish ladders at Stranmillis and then at the weir here, in piers one and three, we have fish passes so fish can go through at any state of the tide. The security man here has seen foxes regularly at night on the security cameras, and three times he has seen an otter, always on the County Down side of the river, on the quay, going round the weir, always going upstream. Then there are the starlings, they used to be under the Queen Elizabeth Bridge and then they moved to the Albert Bridge.

We monitor water quality weekly, twice a week in the summer, we monitor dissolved oxygen, salinity and temperature. Dissolved oxygen is one of the most important things. Because of the brackish nature of the water – you have a tidal environment here – you are never going to get a proper sort of fishery. The water quality is improving, but there is still a way to go. There are a lot of combined sewer overflows.

Use of the river here is quite infrequent at the minute for pleasure craft. One reason for that would be because the harbour below us has always operated as a commercial port. That's changing now, they are talking about a marina in Abercorn Basin. We've been encouraging boating here, we are building six or seven berths just outside the weir to try to encourage pleasure traffic and get them up the river. But apart from that, Derek Booker operates the *Joyce*, a vessel that goes up and down the river, and we have had two enquiries about duck boats and they are coming on line, supposedly within the next one to two months. These are amphibious vehicles, they will do a tourist trip all round the town for an hour and then go into the river at Ravenhill Reach and do a wee circuit for 20 minutes and then go out. The boats are bought, they are in the country getting certified by the MCA, and they hope to have them up and running in the next few months. The MCA is the Marine Coastguard Agency. Then there are the rowing clubs and there have been charity dragon boat races. I think we have got nine rowing clubs all between Governor's Bridge and Stranmillis Weir, and they would operate out of seven boat houses, they would be the main users of the river, they would be up and down at all hours.

We have a compressor house on Stranmillis Embankment, a building with three massive compressors, and basically, each of these compressors is connected to pipework. We would have channels of pipework running all throughout the river, seven lines of pipework, I think it is. They go along the riverbed, in the central channel, the deepest part of the water, and all this pipework is connected to aerator heads which would basically be a tube with a fine membrane on it, covered in small pinprick holes. These let off micro bubbles into the water. A lot of people would think that this is to put oxygen in the water. There may be a certain transfer of oxygen into the water, but the primary function is to mix the saline and the fresh water. At the minute they are on all the time every day because we are doing maintenance on them, we are overhauling them. But once we get that sorted, in the winter they should be only on for maybe an hour or two during the day to keep the system pressurised so they don't bury in the silt. The main usage would be in the summer months, April through to the end of August, you know. If you are walking by the river, you will see a line of bubbles, fizzing away, mixing it all up.[34]

The river revetments, which are stone-pitched slopes along the river, they were put in in the 1920s when the river was canalised and they are in desperate condition, so we are going to repair them over the next three years. Hopefully, it will be a continuous job, like painting the Forth Bridge. Then we are going to dredge the river from top to bottom because it hasn't really been fully dredged since '94 and it's in severe need of maintenance dredging. The Northern Ireland Water Belfast Sewers Project, which is ongoing, will be complete by the time we do the next dredging, so we will hopefully be dredging out the last of the contaminants, and the water quality will skyrocket. Another big project we kicked off in January 2011 is the refurbishment of the Lagan Weir. Each of those rams that operates the gates, they are all being replaced.

When the weir was built in '94, I was deputy river warden, so I have had an association with the weir since it was first built. Deputy river warden was a part-time job, and then I got into diving in '98 through

Dredging the Lagan near the Ormeau Bridge, 2011.

my main job, which was as a civil engineer with CPD – Central Procurement Directorate, formerly Construction Service, formerly Works Service. We did lots of inspections, it was a nice job. Now I'm river manager, but I still keep my hand in with the diving. We would use scuba. Because we only use scuba, anything round the weir that is dangerous, any cavities, we wouldn't be allowed to go into them. But we would go on down, check the gates, check the seals, check for scour, basically any structural faults in the concrete or anything. It is completely black, completely black. You would feel your way. Some days you get maybe a foot of visibility, but most of the time, even if you had a torch, it's like shining it into the fog, the light just bounces back. Sometimes you are better just shutting your eyes and feeling.

Every six months the weir should be checked. There are millions of crabs, wee green crabs, loads of them. See round by the weir, I don't know what it is, but it's absolutely hoaching with them. Like, you put your hand on the riverbed and they would be scattering everywhere, wee green-backed jobs. I don't mind them, they are only wee tiny things.

Overleaf: The Lagan at Stranmillis, 1936. The barge in the distance is by the first lock.

Stranmillis and the first three locks

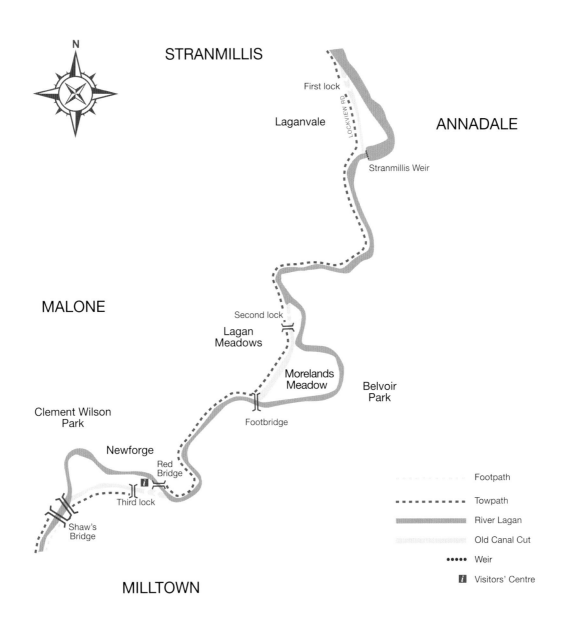

N

STRANMILLIS

First lock

Laganvale

LOCKVIEW RD

ANNADALE

Stranmillis Weir

MALONE

Second lock

Lagan
Meadows

Morelands
Meadow

Belvoir
Park

Clement Wilson
Park

Footbridge

Newforge

Red
Bridge

i

Third lock

Shaw's
Bridge

MILLTOWN

.... Footpath

– – – – – Towpath

River Lagan

Old Canal Cut

•••••• Weir

i Visitors' Centre

Stranmillis and the first three locks
Lagan Lore

At Stranmillis there is a change in the landscape of the Lagan corridor. Downriver there is an urban environment dominated by roads, stone-faced riverbanks and suburban housing. Upriver, the Lagan flows through a more rural setting with grazed meadows, woods and wetlands. The weir at Stranmillis also marks an abrupt change in the waterway from the brackish water of the tidal Lagan to the freshwater river, though walking along the towpath past Lockview Road, perhaps the biggest difference that visitors notice is that the noise of the city quickly fades away. After Stranmillis, traffic is not seen again until Shaw's Bridge.

As early as 1608, it was suggested that if the Lagan was cleared, this would aid the extraction of timber from the Lagan Valley woodlands, and the idea of a canal in the Lagan Valley to link Belfast and Lough Neagh was first proposed in 1637.[35] The value of the river for transport was publicised in 1744, when it was observed that 'by the Help of the Tide' boats could reach Belvoir, a short distance upstream of Stranmillis, and that here the land had been improved by the transport by boat of limestone to make 'lime manure'.[36] Eventually, an act was passed for the creation of a canal and when work got underway in 1754, the *Belfast News-Letter* reported the event with a degree of excitement.

> 'About three Weeks ago they began to work on the Laggan Navigation, at about two Miles from this Town, where the River and the Sea meet; about 200 Men are employed therein, and the Work goes on with great Spirit and Success, being favoured with fine Weather.'[37]

The construction of the canal was entrusted to Thomas Omer, who had previously worked on the Newry Navigation. Some indication of the scale of the project is given by records listing expenditure which detail the employment of staff, including Joseph Wetherby, engineer; Laughlin Byrne, chief carpenter; John Barclay, chief overseer under Mr Omer and Samuel Smith, overseer of the track road. Amongst others who received payments were John Kane, for 'boring and blasting stones', John M'Clean, for 'making a bank and rampart at Ballydraine lock', Elias Clegg, for glazing two lock houses, John Warrick, for gravelling a section of the track road and Cornelius Shean for hinges for gates. There were also bills for the employment of stone cutters, masons, smiths, carpenters, 'drudge-men' and labourers, and sundry expenses that included horse-work, 'curing a man bruised at the canal works' and liquor for men rafting timber.[38]

The canal between Belfast and Lisburn used the bed of the river, straightened in places by digging cuts, the water level controlled by weirs, sluices and locks. The first lock and a canal basin were constructed at Stranmillis (now the car park by Lockview Road) with a second lock at Belvoir and a third at Newforge. A

The first lock at Stranmillis by James Glen Wilson, 1850. The building on the left was Lockview House. Today the track alongside the canal is Lockview Road, the canal basin is now a car park and Cutters Wharf bar and restaurant is situated where the canal lock opened onto the Lagan.

lock-keeper was employed to manage each lock and was provided with a lock house by the canal. The progress of the canal must have excited interest, and when in 1758, the socialite Mrs Delany visited Arthur Hill at Belvoir Park, she wrote in her diary, 'I had the curiosity and courage to go *through a lock!* though I was assured there was no danger in it', adding that Mr Hill and his family 'go almost every day for pleasure'.[39] It took nine years for the canal to be completed as far as Lisburn, after which the pace of construction slowed, and it was only in January 1794 that the canal finally opened to Lough Neagh.[40]

Industries located by the Lagan benefited from the opening of the canal, such as lime kilns at Stranmillis, which in an advertisement of 1779 for the sale of lime by the barrel, noted that the site was 'very convenient for Canal Carriage' and interestingly added that there was also a ford at low tide across the river.[41] Similarly, when Charles Lenon in 1797 offered for sale bricks at his brickyard at Strand-Mills

(Stranmillis) he added that 'he can lay them down any where along the Navigation'.[42] A few decades later the Lagan Brick Manufactory, sited near the first lock, also advertised to supply bricks to 'Gentlemen residing contiguous to the Lagan Navigation' as well as places by the coast between Belfast and Holywood or Whitehouse.[43]

The first lock was an obvious place for the Lagan Navigation to have offices and in the early 1850s, the company secretary James McCleery resided here, at a property facing the canal basin called Lockview House. Nearby, there was also a carpenter's workshop and saw pit, most likely for making items like lock gates.[44] The canal at Stranmillis became the site of industrial development around the end of the century when the Lagan Vale Brick and Terra Cotta Works Limited took over an existing brickworks and laid out sites for housing and factories.[45] Businesses were encouraged through advertisements that promoted good access to the Lagan Canal and an abundance of water for manufacturing. In the following decades a surprisingly diverse range of industries were to locate at Stranmillis, particularly along Lockview Road, including Vulcanite which manufactured roofing felt. This industrial process required large quantities of water which were pumped from the Lagan and fine sand which was brought from Lough Neagh in lighters owned by the company.[46]

Today there are few signs of the industrial heritage of Stranmillis. All of the tall factory chimneys have gone, the brickfields where clay was extracted have been developed for housing, and almost all of the factory buildings have been replaced by apartment blocks. However, the reuse of land has helped ensure that Lockview Road still marks the limit of suburban Belfast, and further upstream the river corridor remains a largely pastoral landscape. These lands include the Lagan Meadows, which Belfast Parks manage by low level grazing, and associated wet grasslands that have for many years been an Ulster Wildlife Trust nature reserve. A network of paths provides public access to this area, which is of particular importance because of its wildflowers. One path along the northern margin of the wetlands passes Leister's Dam, which was originally constructed to provide a water supply for Belfast, though today it is largely drained of water.[47]

In this region steep slopes near the Lagan form attractive undulating landscapes, and amongst the trees on the County Down bank, an impressive mansion was built in the early eighteenth century at Belvoir Park. Smaller mansions were constructed nearby at Annadale and at Deramore (now Ardnavally), and on the County Antrim bank there was a large riverside house at Newforge. Following the First World War the impact of social change, land acts and taxes, including death duties, caused a decline in the fortunes of big houses, and by the 1920s, the mansion at Belvoir was empty. A new use was found for this estate during the Second World War, when the mansion house was occupied by military personnel, and the grounds became an armaments depot: ammunition was brought in barges from the docks and unloaded at a wharf constructed on the Lagan.[48]

Unfortunately the mansion at Belvoir and also the houses at Annadale, Deramore and Newforge were all eventually demolished, though much of the former grounds of Belvoir remains as a forestry plantation with public access. Here one can walk along old estate paths, explore the former pleasure gardens and, if you are lucky, still see a red squirrel, though numbers at Belvoir, their last refuge in the Belfast region, are rapidly declining. Along the river can be spotted kingfisher, otter and even occasionally a seal. Recently, a most remarkable discovery was also made in the woods at Belvoir Park when a number of fallen oak trees were tree-ring dated. Several were found to be over 300 years old with one having started to grow in 1642, the oldest date found so far for any tree in a woodland in Ireland.[49]

Between Stranmillis and Shaw's Bridge there was an industrial area at Newforge where, as the name suggests, there was a forge. This was one of a number of ironworks by the Lagan in the early seventeenth century that exploited local woodlands to make charcoal to smelt iron, and it is likely that they used the Lagan to transport materials.[50] The ironworks at Newforge was destroyed during the 1641 Rebellion, though by the 1760s, there was a bleach green in this area and also, by the 1880s, a beetling mill powered by a waterwheel.[51]

The Newforge mill was one of many that took water from the Lagan, and a court case in 1884 provides fascinating insights into the conflicting interests of mill owner and lighterman. On 20 September 1884, over 40 lighters were stuck near Shaw's Bridge because of a lack of water in the canal. Although other factories cooperated to raise the water level, and a flood of water was beginning to come down the channel, the owner of the Newforge mill decided not only to continue to run his factory, but to continue to run off water. Under the gaze of the canal manager, who was surveying the scene from Shaw's Bridge, a canal boat moved to block the mill race, the mill owner was knocked to the ground and kicked, and machinery in the mill was broken. When the case came to court, the mill owner claimed that running off water was just his usual practice to clean out the mill race and that he had not noticed the stranded barges. A lighterman was found guilty of assault though it would appear that the mill owner did provoke the situation, as it was revealed that in the past the Newforge mill had received payments from the Lagan Navigation Company not to draw off water when the level in the canal was low, but that no such arrangement had currently been in place.[52]

By 1918, a new business, the Lagan Clog Works, was operating at Newforge.[53] This only lasted for a few years, and the site was subsequently taken over by Irish Food Products, a business developed by Clement Wilson. Initially, operations at the plant included rendering, though this gave rise to complaints from Malone residents. Later the factory produced tinned soups and other processed foods. Land around the factory was used to grow some produce, and the grounds were enhanced with extensive gardens that were open to the public. Subsequently, the factory changed ownership and became used for the production of pet food. Today the industrial buildings have gone, replaced by offices, though the former gardens and open space created around the factory have been preserved as Clement Wilson Park.[54]

Stranmillis Weir, 2011.

Swimming used to be popular in the relatively clean waters of the Lagan around Stranmillis, particularly an area known as 'the Hole' (most likely the secluded spot just below the weir). A walk along the towpath from Stranmillis was also a popular escape from the hustle and bustle of urban life, though around the latter part of the nineteenth century, there were letters to the papers complaining about what one correspondent termed 'unconventional' bathers. The author of a letter of 1875 exaggerated perhaps a little in expressing concerns that because of youths swimming without costumes, 'Ladies could not possibly have walked along the path, and those in boats could not have got back to the landing-place'.[55]

The towpath around Stranmillis and the first lock is often referred to as 'Molly Ward's' though who exactly Molly was seems to have changed with each telling of her story. In the lovely book *As I Roved Out: A Book of The North,* published in 1946, Cathal O'Byrne refers to Molly as having been an innkeeper who lived by the first lock, where there are now tennis courts and boat clubs. O'Byrne describes the tavern during the eighteenth century, at the time of the United Irishmen when powder and arms were hidden at Molly Ward's, and how Molly thwarted the best efforts of an aptly named Major Fox to find the weapons.[56]

Molly Ward's Tavern between the first and second locks, 1888.

Like many a good story it is difficult to know if it is based on fact, although another book, published in 1866, a compendium of confusing snippets entitled *Rambles and Gossip along Highways and Bye-ways round Belfast*, provides some further information. The author, Francis Davis, asserts that, while Molly lived in a cottage by the first lock, she was never an innkeeper and that her husband had been a night watchman when the canal was being constructed. This information Davis seems to have obtained from a Mr John Ward, whom he describes as the son of Molly, and who was then living by the Lagan.[57]

During much of the nineteenth century there was, however, a tavern in the area. This was a short distance upriver from the first lock, a hostelry by the towpath that was variously known as Ward's Tavern, the Lagan Tavern or Molly Ward's. It was enlarged in the early 1850s,[58] and when, in 1857, large crowds came to see an Easter firework display devised by one Professor Johnston, the establishment was described as 'that favourite resort of the working classes'.[59] By the late 1870s Ellen Keely was the proprietor and when she in turn sold the tavern it was run by the McAuley family, who also had a horse for drawing lighters on the canal.[60]

Despite the popularity of this riverside establishment, it eventually shut its doors for the last time. Most likely it lost its licence – the authorities may have had difficulty controlling a pub that was situated on the Belfast boundary and accessed on foot or by boat. The outcome of a court appearance by Ellen Keely in 1882 would also have been prejudicial. Ellen, who had a six-day licence, was accused of selling intoxicating drink on a Sunday. In her defence she claimed that a dog had strayed through the open door to the tavern and that two men had followed it in.[61]

Although the story of Molly Ward is now half forgotten, and all trace of the tavern has long gone, there are many people who recall resting awhile at another riverside halt at the third lock run by George Kilpatrick. George, who had been the lock-keeper since the mid-1920s, enjoyed having a chat sitting on a bench at his shop, a lean-to timber structure with a felt roof built against the gable wall of the lock house. Here he sold sweets from glass bottles with big screw-top lids, chocolate, cigarettes and soft drinks. By the 1950s, when traffic on the Lagan Canal was declining and George's family was also growing up, the shop was shut, though George and his wife Sarah continued to live by the lock, and one of their ten children, their son Stanley, stayed on at the cottage until 1993.[62]

Stories of the Kilpatricks' life by the Lagan and of the former shop have played a vital part in the recent development of a small visitors' centre by the third lock, with offices for the Lagan Valley Regional Park, a display area and coffee shop. The centre opened in 2009 and is now encouraging a new generation of walkers to follow the path by the Lagan, explore the region, look around the Kilpatricks' refurbished lock house and see George Kilpatrick's lock, now restored to working order.

We used to give the horse a bit of a lift up the road
Albert Allen

Well, the time I was at school there was nothing only barges coming down, hauled by the horses. The lock-keeper down there at Stranmillis was Billy Gratten, William Gratten, and he lived in the house – demolished now – by the first lock. And he had three daughters: Natasha, Martha and I forgot what they called the other girl – Isa. He had been an old naval man in his day and he was very rough! He done a bit of business letting out rowing boats, I think he charged six pence an hour. But he was a bit rough, and if we were down fishing at all he was after us, you know. In case you would fall in. He was at the first locks, Billy Gratten. We knew him well, but he wouldn't have let you after the tiddlers in case you fell in. And he cursed a lot, you know, and he used to chase us away, but there was hundreds of barges going through the lock at that time.

I remember coming home from school and the old tug was towing up, I think, about 18 barges, and that was at three o'clock. At half past three he was putting them in through the lock and he was still at it at eleven o'clock that night, putting them through. The horses was waiting to hitch onto them and start the towing. That would have been about 1918. I was born 1911. They took coal through and Indian wheat, and you want to see the rats after the Indian wheat; they used to be scaling the ropes, and we used to watch them scaling the ropes to get in. And we had a wee collie used to kill them galore! They used to take the coal there to John Shaw Brown's, and I think they did Barbour's too and they went the whole road across Lough Neagh, some of the barges. There used to be a motor barge, a diesel one, they called it *Industrial Lambeg*, it was run by two brothers, they called them Spinlow that lived up the road. And they usually carried wheat and coal, too, and sometimes a load of salt, they went down to the salt mines at Carrick and loaded on there. And I don't know where they took it after that. The two brothers lived up in Stranmillis Street and one of them was a boozer and the other one was in the Salvation Army!

But Billy Gratten was a rough old fella to deal with. You would have heard Billy Gratten shouting from midnight getting the boats through, and they were always arguing who went through first. And I remember one time someone threw a big collie dog into the lock, and it was swimming up and down, it couldn't get out, and my brother and me didn't know what to do, to bring Billy Gratten down to get it – he was afraid to go near it and so was I. My brother went up and seen Billy Gratten and told him about the dog and right enough he came down and got the dog out. Someone threw it in to drown it. They did.

But I was more in conversation with Geordie Kilpatrick at the third lock than I ever got in conversation with Billy Gratten. And then there was a fellow at the first lock with Billy Gratten, they called him Dick Belshaw. He was the weir man. You ever heard of him? Dick Belshaw. And he used to get drunk too! His job was to regulate the water and keep the banks from flooding by drawing up the weir. And there was a man who used to do the banks, I was very friendly with him, they called him Jack Jones, and him and his wife lived in one of the cottages there. And he went up one time, and you had to put this old iron bar into the sluice gate to higher it up, you know. And it was in the month of August, and it stuck and he couldn't get it down, and with this old iron bar he made a swipe at it and him and the iron bar fell into the river. And just as he fell in, the weir slid down and jammed him. He was drowned. The fella you called Jones. I knew him well.[63]

My brother worked in the brickworks at Stranmillis – he was a burner, you know. When they put the bricks in soft into the kiln, his job was to burn them to harden them up. And then at that time there was a brickworks across the river, I think they called it Annadale. I knew men worked in it. They used to load the brick on. To take them different places. You will see it near Annadale School where they used to load the brick onto barges, there is a kind of a wall there.

They used a kind of breeze coke to burn the bricks. They had a boiler there, my brother was one of the boiler men. And I think they used to use it or else coal to burn the brick. Took about a fortnight to dry them out, you know. Then sometimes if the weather was good they used to put the brick out in what they called hacks. Like a straw thing, you know, and they put them out for the sun to dry them. And then these hacks was put on top of the bricks to save the rain getting at them. They nearly depended on the air drying them out. Then I think they took them down and burned them down there, and then it didn't take them as long. My brother was in it for years as a burner. But sure they only paid about 30 shillings a week.

And then there was the Vulcanite. It got the name of being a slave house. Facing where Cutters Wharf is now, there used to be a laundry there, and the women used to do the washing and all, the drying for some of the laundries, but it was just fading out when I came on the scene.

The brickworks boiler house? The one my brother was at was just an ordinary boiler. He used to shift the coal in, rake the ashes out, fill it again with coke or coal. And then it kept it going. But it was hard work. The old manager wouldn't have paid them too much and I think the Vulcanite was worse. You see, what they done was the raw clay was dug out of the bank and then it was put in a great big heap, you know. Maybe half the length of this street. And it had to get weathered for about a month, and they used to have a fellow in it for picking the wee stones out of it in case it damaged the machinery. And it had to weather. Then after it weathered and settled it was dug and put on a wagon and took down and put in to be made into brick. The wintertime, they yearly dug the raw clay out and weathered it over winter and then in the summertime it was taken down and made into bricks. And they couldn't work if it rained, for the clay was too wet. They couldn't work.

But I remember, too, one Sunday evening there was a place away down near Ormeau Road let boats out, pleasure boats, and these three girls and three fellas got into this boat down at Ormeau Road and

The Lagan looking upstream from Stranmillis, 1929. A barge proceeds upriver towards the first lock which is visible in the distance on the near bank. Also on the near bank are the tall chimneys of industrial buildings along Lockview Road, including in the foreground Edwards' soup works. On the far bank are the chimneys and industrial buildings of brickworks that extended from Ormeau to Annadale. The roads along both riverbanks had recently been constructed as part of the Boulevards scheme.

rowed up behind the weir and they started fooling about, rocking the boat, and the six of them was threw in and the six of them was drowned. And Mr Phillips said – he was an old navy man, he worked for the boat club – and he said he could have got them all out if they hadn't have panicked. He never got even one of them out. They all lived about the Ormeau Road and I think there was one or two good swimmers, and he said he had one of the swimmers out but he went in again after one of the girls and she got the hold of him and the six of them was drowned. That was in the month of August.

I remember the men all saying that it was very hard work in the Vulcanite, and the man who was manager in it, you called him Donnelly, Robert Donnelly, and he was all against them joining the union, and there was never anything else only disputes in it, but they finally got it settled and they joined the union. There was women and men worked in it, they made felt and they made tar bitumen for roofs, you know. They made that. But they seemed to do rightly. I heard one or two that worked in it said the money was poor and the work was dirty and hard.

'But it was hard work. The old manager wouldn't have paid them too much.' This photograph of a horse-drawn wagon at the Lagan Vale brickworks probably dates from the time of Albert's youth.

Then, you know where them new flats are, down there, them new apartments? That was the soup works. It came to the fore by making soup in the 1914 war. They made money shipping it for the troops. You don't remember it? The soup works. The horn blew at six o'clock for them to start, then it blew again at eight o'clock for the others to start, and the brickyard horn went off at half seven and the Vulcanite horn went off at eight o'clock! And some of the people said, 'I wish them ol' horns would stop blowing!'

Edwards' Soups, I think they are still on the go.[64] And they also used to have pickled onions. It was mainly potatoes and spices and things. Dried out and ground into a powder, you know. And then you cooked it to make soup. They say they done well during the war for it was shipped to the forces. Just powder, more like they brought it down to small granules, you know. It closed after the war. A fellow there called Schofield run it. He was the last and it closed down. He went to live in Newcastle. And manys a time when they were taking a load of soup in a horse and cart up the road on a frosty morning, the horse couldn't make it, and the old master got us out of school, about 30 of us, to give the horse a push up the

Right: Poling a barge at Stranmillis. On the far bank is the Belfast Boat Club, which was destroyed in a bomb attack in the 1970s. The canal basin is now infilled.

hill! Manys a time it happened. We used to give the horse a bit of a lift up the road. The school there was just up a little bit on the right-hand side, up Stranmillis Road before you come to Ridgeway Street.

Oh, but the old school life was a rough life in them days. Never liked school. I hated it. Ol' masters were doing nothing only bothering the children. There wasn't that many schools about at that time. And I pleaded with my mother to send us somewhere else, but she wouldn't do it and the old master was an old taskmaster. He beat the life out of the pupils. They wouldn't stand for it now in the schools, you daren't touch them now. And then the schools were very cold, too, in them days. And the winters were harsh.

Then old Mickey Taylor. He was a good age. And Doctor Moneypenny said one time to my mother, 'Mickey Taylor will not last the night'. He said, 'He is on the way out, he will not last the night.' And the next day I seen him wheel the barrow down the street! I think he was in – they said Mickey Taylor was in – the Crimean War. And he was in the Boer War, too. He was a man about five foot ten and very thin, very thin when I knew him. Of course, he was near 100 years old. He lived in Wansbeck Street down there after he gave up the job at the locks.[65]

But at that time all around here was very rural. Down Lucerne there was a wee river flowed. And we used to fish for the tiddlers down Lucerne, it went right down and into the Lagan. It's diverted now. And the hares used to run here. And then there was the first Malone Golf Club was there too, at Richmond Park. The old club house and the course used to go round down near the River Lagan and home by Knightsbridge. And they teed off from Richmond Park. Then round here there was nothing, only cattle grazed. Up at the back of our garden they milked the cows. And there was a man there lived in a house there on the Ormeau Road, off Sunnyside Street, they called it Prospect House. Well then, the old fella who lived in Prospect House, he used to milk the cattle up where our garden is now. He was totally blind, the sons had to lead him about until he started milking. Stevenson they called him. They lived in Prospect House, it was a big mansion house in its day, you know. And then Sunnyside Street, at that time there wasn't a house in it. It was occupied by the gypsies. Nothing only gypsies in it. And they were fighting all the time.

At that time there was no embankment whatsoever. If you wanted to go to Sunnyside Street from Lockview, you had to go up the road and down Ridgeway Street. Over King's Bridge. They started to build the embankments when I was at school. I was about 12 years of age at that time, that must be 80 years ago they started to build it. And you know what they built it on? Food chits. There were no wages. You got a ticket to go and buy your groceries or your clothes or anything. They called them food chits and clothes chits. That's what the embankment was built on. They built it shovel and spade and an ol' roller. We used to watch them from the school grounds. They built the one from the roundabout 'til King's Bridge first, and then they went across and built the one up to Annadale. And the one from the King's Bridge to Ormeau Bridge and the one from King's Bridge to Ormeau Road, they were the last two built.

Harleston Street was known as Piano Row
Mervyn Patterson

I was born in a terraced house in 5 Laganvale Street on 17 June 1925. My father came to Stranmillis from Stewartstown, County Tyrone, sometime early in the century to work in a new bottle works, a glass bottle factory establishment situated approximately where the Ordnance Survey is today, their offices on Stranmillis Road.[66] I am not sure what his function was in the bottle works. I think perhaps he fired boilers or something like that, but I am not sure. They definitely blew glass because I remember we had two decorative glass pistols my father had blown.

Number 8 Harleston Street became available. Now, in Laganvale Street they were a front bedroom, back bedroom and then what was known as a return bedroom, a two-storey return block. Though there were three bedrooms in 5 Laganvale Street, one had to go through the bathroom to go into the return bedroom. Well, this house became available in Harleston Street. Front and back bedroom were on the first floor. On the second floor there was an attic bedroom and a bathroom containing only a bath. Each of these rooms had a roof light, a skylight window. Each house had a small forecourt garden and an enclosed yard at the rear where an outside WC and fuel store were located. The attraction of this house was that, for a big family, you had the attic room where you could put two double beds. I lived in this house until I was married in 1954.

Mervyn Patterson in the lane behind his home in Harleston Street.

We called it Laganvale Village and there were names for the streets apart from their official title. For instance Harleston Street was known as Piano Row, Laganvale Street was known as Tar Row or Tower Row because of its flat roofs and Wansbeck Street was known as Poverty Row.[67]

The lamplighter was Thomas Pinkerton. He lived in Coolfin Street on the Donegall Road and was lamplighter for the Stranmillis area. He came in the morning and evening to be sure the lights were off and on. During the day he cleaned the glass, replaced mantles and adjusted the time clocks. He rode a bicycle with a ladder over his shoulder, but at night he had a chain with a thing like a footrest attached that he put round the lamp standard – he could stand on this to reach the light.

Flat roofed houses in Laganvale Street, 2011.

In my time, when I was young, there were four shops in the village. Development took place between the streets in the late '20s and early '30s along the frontage of Lockview Road. You will see they are roughcast houses, different from the red brick houses. Well, there was a lady there who had a shop in the second of those houses. Doherty was their name. Doherty's shop sold mainly confectionery, cigarettes and fireworks, that sort of thing. But that didn't take place until about 1930. Well then, in Laganvale Street there were people called Ingram, they would have lived in about number 27 or 29, somewhere there. They were a family that came over from the glass people, Pilkingtons, in the north of England, to set up the bottle works. They had a little confectionery shop there. Further up the street, oh, it would be about number 33 or 35, something like that, people called Slynn – again they came over with the glass people – they ran a fish and chip shop there, believe it or not. The grandfather of the Slynns – he used to have a cart with two bicycle wheels on it and he sold fire lighting sticks. It was just a terrace house, what was their reception room was the actual shop. Well then, in the last house in Laganvale Street, number 43, there was a Mrs Wright lived there. Again, one of the Pilkington crowd, and she made a very refreshing drink. It was a dark brown come orange colour. This was known as pop. The people of the village tried to watch her to see what ingredients she put in it, but she never let anybody know that! There was a man, Thomas Neill, lived in about number, what, 35 Laganvale Street; Mrs Wright lived about 43, and when Mrs Wright died, Tommy Neill's lament was she died without leaving the recipe for the pop! I think it was a penny or tuppence for a bottle, you usually got it in a bottle with a spring clip lid. One just called at the front door.[68]

One other thing I should tell you about Laganvale Street – it would be about number 25 or thereabouts. This house had a single-storey return block there and on top of the return block was a wooden room and that was the original band room for Laganvale Flute Band. It was a part flute band, it wasn't a blood and thunder band, and just before the outbreak of the Second World War, 1937 or thereabouts, it went over to be a silver band. During the Second World War Laganvale Band was affiliated to the ATC, Air Training Corps. It is still about, it is known as a concert band now. Their hall now, in Wansbeck Street, it used to be a wooden structure, but some years ago it was rebuilt in concrete blocks.

In Wansbeck Street there was a shop at number 1, a man called Bobby Callaghan occupied it in my time and then people called Nesbitt took over followed by their son-in-law called Cole. I think the shop window is still in existence, though it's all a dwelling house now. It was a general grocery store, confectionery and cigarettes. Well then, on the opposite corner of Wansbeck Street, and fronting on Lockview Road, there is a place there, I think it is AA Components or something like that now. That was a grocery and confectionery shop. It was run by a William Houston, I understand he came from Killinchy in his day. Well he marketed groceries of all sorts, made his own ice cream. Eggs in it and natural cream and so on, it was wonderful stuff. Indeed from the Commercial Boat Club down at Balfour Avenue, you could hire boats there, and boys used to row their girlfriends up and beach the boat and go and get an ice cream at Houston's or whatever. Well then, Houston, he kept almost everything and he had stabling out the back and he stored and sold hay. The lightermen – the men who drove the horses which towed the lighters – used to stay in the stone-built and slated stable where Queen's University Boat Club is now. It had about four stalls in it, but if it was oversubscribed, if all the stalls were full, some of these lightermen could go up to Houston's and hire a stall out from him for the night. And he would sell them hay and corn as well. At the back of the stables at the rear of Houston's shop – I think it's still there – there is a wooden structure, a wooden building. I don't remember it being used as such, but my older brothers told me it was used as a dance hall, it was known as the Pioneer Dance Hall. Dances were run in there. It has been used as a paint store in more recent times.[69]

Next to the bottle works where my father worked was the Irish Cold Bitumen factory. They laid bitumen on pathways and that sort of thing. You used to see tar flowing down the bank at the back of the factory towards Stranmillis Embankment; they must have had a leak or something.

Telephone:-
BELFAST 65017.

Telegraphic Address:-
"DESICCO, BELFAST."

Fredk: King & Co. Ltd.

E·D·S

Sole Manufacturers of

EDWARDS' DESICCATED SOUP

REGD TRADE MARK.

EDWARDS' DESICCATED SOUPS.

STRANMILLIS EMBANKMENT,

BELFAST.

6th May, 1957.

Letter heading for the soup works.

On Lockview Road on the left was Edwards' soup works, they made desiccated soup from potatoes and onions and spices. Apparently, they put this mix on trays in the ovens and it came out as brown pieces, mid brown in colour, rough pieces of stuff. When the men were coming out off the night shift, they would have some in their pockets and they would give us lumps when we were going to school. It was very hard but it would soften when you put it in your mouth. It tasted a bit like Bovril. They also marketed vinegar and spices. A late uncle of mine who worked in the soup works after the war used to be able to get large bottles of Flag Sauce at a very reasonable price.

On the right, at the commencement of Lockview Road, where there is now a row of shops, there used to be a big house on a large site. I think it was a red sandstone building, it was like something you would see on television, a haunted house! It was called Ash Grove and was where the manager of the brickworks lived. In my time a man called Vaughan lived in it and the corner of Lockview Road and the Stranmillis Road was known as Vaughan's Corner. Vaughan was succeeded as manager of the brickworks by T Courtland Hunter, who resided in a short, narrow cul-de-sac off nearby Richmond Park.

Going further along Lockview Road was Campbell Brown's foundry. They perfected a machine for cutting soda farls! And they patented that and I know the Ormeau Bakery used that for years. Adjacent to that, long since gone, there was a single-storey stone-built building, slated roof. I understand it was a laundry. It wasn't so used in my time, but my older brothers often talked about it. It was later used as offices by the nearby Campbell Brown's foundry.

The next commercial establishment was the Lagan Vale Estate Brick and Terra Cotta Works. They made bricks and ornamental terra cotta facings and decorations for all sorts of properties. Indeed, many of the rich brown bricks you see in buildings in Belfast came from the Lagan Vale brickworks. They had a four-wheeled flat cart for delivering bricks that was pulled by a white horse called Billy.

Well then, above the Lagan Vale brickworks on a private road, now the continuation of Lockview Road, was the Vulcanite factory. It manufactured roofing felt. There was a large house by the Vulcanite factory where the manager lived, a man called Robert Donnelly in my time. Another house adjoining the Vulcanite offices was occupied by the firm's engineer, a Mr Matchett.[70]

On Lockview Road beside the brickworks offices was a double-fronted two-storey detached house where two ladies known locally as the Heavenly Twins lived. I heard that they were the Miss Williamsons. The two sisters and a brother were, I think, from Cavan and they had to leave in the 1922 Troubles. They always came out in 1880s style dress, long crinoline dresses and always wore large hats.

Up Sunnyside Street at Ballynafeigh there was a firm called Stevensons. The senior man there, the old grandfather, I remember him. He was blind and they had cattle grazing over the fields in the Lagan Valley, and I do remember them up tying the cattle to the fence and helped this man over 'til he got his hand on the cow. He had a three-legged stool, he would sit down and milk away. It was known as the Prospect Dairy. They milked and delivered milk. They grazed just from beyond the Vulcanite factory, all those fields there.

You probably know there is a pond shown on the Ordnance Survey map, Leister's Dam, Belfast's first water supply. In my time the water just ran away there. But on the bank of the hill at the back of the pond – we knew it as Sloan's Pond, someone called Sloan must have lived there – there were fruit trees and all, but I never remember a house there in my time. But anyway, a man called Dixon who worked in the nearby YMCA sports ground and wore a patch over one eye, he put a piece of wide concrete pipe over the spring at the back of the pond, it sat up above the surface of the land. He put a barbed wire and wooden post fence round it to exclude cattle and had large stone steps down to it. There was a cast iron drinking cup and at the time of the Blitz you could go there for a drink of pure water.

If you walk along the towpath of the Lagan today, you will see along the riverbank trees, bushes, all sorts of things growing. But they would have snagged the tow rope. The bank ranger, his job was to keep trees away, cut them down. The hedge on the other side of the towpath, he kept that trimmed and the towpath is now nicely tarmacked, but it was just rough hard core and holes would appear in it. He would have filled these. There is another thing: the Lagan Navigation Company had a squad of men working for them as well as the lock-keepers and bank rangers. And they had a large flat-bottomed craft, it was known as a scow. And they came down and carried out repairs to the bank and so on.

The first locks on the Lagan Navigation system were beside Laganvale Village. The lighters carried maize and coal, et cetera. If one went down and assisted the lighters through the locks by opening and closing the lock gates, one was allowed to pocket a few handfuls of maize which helped to feed hens which some people in the village kept. My father kept 12 Rhode Island Reds in the backyard of our house. The world was a very different place in the 1930s.

The Belfast Boat Club is beside the weir and opposite the site of the Vulcanite factory. The canal, now filled in and used as a car park, flowed in front of the clubhouse. The boat club was an extremely posh establishment and was reached by way of a flat-bottomed punt with a rope at each end, and schoolboys from the village earned a few tips by pulling the punt to and fro. They had a tin box sitting on the punt, and the gentry would drop a thrupenny bit or coppers in. At tournament times this was quite remunerative. It was very upmarket and only the merchant princes and very well-off could obtain membership. I remember hearing that the twin daughters of Donnelly, the Vulcanite manager, were considered unsuitable for membership despite their brother being a medical doctor with a GP practice in Donegall Pass.

Boats were hired out at Commercial Boat Club at Balfour Avenue on the Ormeau Road and people rowed up. They rowed up to the back of the weir at Stranmillis. In fact, in that backwater, downriver a bit from the weir, there was a hulk of an old ship there. I don't know how it came to be there. There was a man called Girvan lived in about number 41 Lockview Road, he had some sort of a job down at the harbour, I think he was in a supervisory position, he brought a squad up and they sawed up and took that old boat away. But it was there for donkey's years.

Well, then, we all learnt to swim in the salt water part of the Lagan. Indeed, Wellington Swimming Club used to run an annual race – you dived in off the wall at the first locks and swam to Commercial Boat Club at Balfour Avenue, which is about the third street on the right going to the city from the Ormeau Bridge. It was a local swimming club, they swam in Ormeau Baths, but they ran this race annually and it was great, spectators lined along the Boulevards. It was a big thing in those days. I am talking about just after the war, late '40s, early '50s. But then for some reason or other it was discontinued. I entered it. I joined Albert Foundry Swimming Club. I could have swum for a week but I was one of the early ones in because I wasn't a speedy swimmer. And all these fast swimmers come down after they got in, flying past, water flying everywhere. But when I got about the King's Bridge they were sitting in boats, they had got tired, I always made it.

Not being allowed to go up there, that's where you went
Ernie Andrews

At Annadale there were two quarries, one in use when I was a kid, the other disused. The disused one was the one we would have been in, because it was full of newts, it was flooded. Little streams ran in through it and over the very soggy ground, and that used to build up the water and we would have got jam jars and away over for newts, any God's amount of them, and took them home and immediately got sent back with them! 'Take those back where you got them from!'

The quarry that was in use, I can remember men working down in it, but we really didn't go over too close to it, we weren't really allowed near it. The brickworks itself and kilns were on top of a hill, and the clay was taken up with a little railway, a little very, very steep railway, it must have been operated by cables that took the little trucks on up. At the bottom of that hill near the railway were two ponds, one was Martin's Dam – we knew it as Marty's Dam – and the other was known as the Cooler. The water was taken, from what I can remember, from Martin's Dam up to the kilns. They must have sprayed the bricks to cool them and then the water that was left ran down a pipe and into the Cooler. And we always said that

Ernie Andrews with Buddy, late 1990s.

the newts and frogs and all in the Cooler were much bigger than the ones in Marty's Dam because they lived in hotter water! But being kids we were not allowed over there, officially we were warned not to go near it by our parents but of course, that was somewhere you had to go. This was in the early 1950s. But my grandfather was brave and sharp, an ex-army man from the Boer War, and he could have told where I had been by looking at the soles of my shoes, the clay. So being sharp kids we tore sacks up, old sacks, and we actually tied them round our feet! And we could go over into the brick fields and round to Marty's Dam. And back onto Sunnyside Street when you were heading back home you would have taken the old bits of sack off. And my grandfather would have inspected your feet and, 'No clay there, no, he is all right, he's not guilty!' So we got away with it for years![71]

The old disused quarry was right over beside Annadale School where it is all new houses now. Deramore Gardens would have been a dead end but if you came to the dead end and looked down, you were looking into the old quarry. The one that was in use was over towards Sunnyside Street. Now the houses that are up, the highest houses of the older houses, they are up where the brickworks was. Is it Ailesbury Crescent? I think it was up in there. If you went to the top of Kimberley Drive and straight on, that was where the brickworks was. There was a big chimney, massive big chimney, a big factory chimney, I remember it coming down, I remember them knocking it down. Oh, that was, it must have been late '50s. Surely it must have been around that.

And a dairy, I am almost certain Dorman's Dairy, was at the top of Kimberley Drive where you would have got to the top of Kimberley Drive and turned right. I remember hens and all knocking about there. That was another way into the brick fields. Also from Sunnyside Street, from Highlands shop. Just below the shops was I think a glass factory because we found loads of glass and stuff, though if there was a factory this was gone by the time that I was knocking about. I heard there was one on the Stranmillis Embankment, over where the Lyric Theatre is. There was one there, I don't remember much about it but I have spoken to people about it who said, 'Oh yea, there was a glass factory at Stranmillis'.

Well, you got to the bottom of Sunnyside Street, turned into Annadale Embankment, prefabs were right along. There was a whole little estate of prefabs in there. Immediately after them – which would be roughly level with where the Governor's Bridge is now – was waste ground. After that was the start of the plots, the allotments, and they ran the whole way up Annadale Avenue past Annadale School right up to the bad bend. And behind those plots, that was the brickworks, that was the whole area of the brickworks. And whenever the brickworks was done away with, the plots and all went, prefabs went and a new housing estate was built, which was known locally as the electric houses, because there was electric fires and stuff in them. The plots at the bad bend remained up there and are still there. Across the road, at the edge of the Annadale dump were the new plots that my grandfather and his friends were moved to, and the ground was terrible, they were just a failure. They couldn't work with the ground at all and they lost heart in it. The electric houses had no chimneys; they were the first that we had seen that were all electric, no open fires. The prefabs were detached, all detached, there was a block of them right at the bottom of Sunnyside Street, on both sides of the road. There were also more over the river around the bottom of Ridgeway Street. They were built I think immediately after the war.

Further on up at the Annadale dump at the bad bend by the river, we would have spent a lot of time up there. Again, not being allowed to go up there, that's where you went. My grandfather's plot was up there so I used to go up with him on a Sunday – my mother was glad to see the end of me, get a bit of peace on a Sunday, so I went up there with him. And he would have took me down to the dump for a look, and I remember the rats. They were big! Many of them. But there were also half-wild cats that lived in the

Ernie Andrews with his grandfather, Robert Short, and his mother's sister at the Annadale allotments, mid-1950s. The tall chimney in the distance stood at one of the Lockview Road factories.

rubble that had litters of kittens which never did well – they had bleary eyes and some of them had only three legs. But the dump used to go on fire. Every now and again it went up, and the rats would have evacuated and whenever they decided to make their move nobody hung about! Because they went down Annadale Avenue in dozens to get away from the fire. The firemen used to come up and try and damp it all down, they used to take water from the Lagan, they used to pump it up. They were regularly at it, I don't know whether the dump went up itself or somebody was setting it on fire, but up it went. I remember the wee bin lorries going down, the wee ones, they were sort of rounded shapes and they slid the doors up on them, they tipped them. And there was an old boy that was like a watchman, he kept an eye on things because his job was not only to make sure the bin lorries were going where they should, but to keep the like of me and me mates away, which he had a lot of trouble doing!

You couldn't have a council dump like that now. It was so close to the river, too, I am sure an awful lot of seepage would have went down in at the weir but still that river was very, very clean in those days. And what now has turned out to be salmon in the river were always there, but we didn't know, when we were kids there would be these fish trying to jump the weir at Annadale. We didn't know what they were, these big silver fish, but now with having many seasons of fishing behind me, I know. Oh my God, we had salmon on our doorstep and we didn't know! We didn't know what they were!

In the allotments were all old ex-servicemen that were given these plots, all these old boys. All sorts. They all knew one another. Now it's a hobby, I think, with people, but in those days it was a bit more than that because you ate what was grown up there, I don't think it would maybe be the same now. But money not being as handy in those days a lot of boys sort of made a living from selling their stuff. I don't see the same going on – I would walk up there with the dog, it's more flowers and people just, you know, a break from work and they have got an allotment and they can work in it, which is nice, but I don't think there would be the same urgency to grow stuff now as then. And he grew, I remember, peas and gooseberries, things like that. You learned the hard way about what you could eat and what you couldn't eat – either being sick or hit over the head, some one of the two! He was up, certainly all day on a Sunday you would have got him up there and during the week he would have gone up, he had an old bicycle and would have went up. And I remember the big water barrel and the hut with all the spiders in it, all his spades and rakes and what have you all in it. I always tried to get out of him about the war, about the Boer War and all, but he didn't ever talk about it. Typical of an old soldier, you know, he had seen enough, didn't want to go back to it.

Over on the other side, at Stranmillis, there was a soup works on the corner there, just at the roundabout, opposite the Stranmillis Teacher Training College. That was a big factory, a big chimney there. And they had a very good orchard at the back of it, so we would have frequented it! It was where those new apartments are all built now, you know, beside the boathouse, that was the big soup works. There was no right of way along there at the back of it, but we actually came over the wall there, by the river, until one of us had a bit of a bad incident with a bulldog that was in the orchard and we didn't all get out in time, we almost got out but not quite! So it was stroked off the list of places to go! It was a big orchard, well tended. I think it was perhaps some supervisor from the soup works had it. Maybe it was there before the soup works and they just kept it.

Over near the Teacher Training College at Stranmillis – some of my wee mates were from that end of the world – there are two cracker ponds in the grounds of the training college, and whenever Sir Eric Ashby was the vice-chancellor of Queen's University he apparently got them stocked with quite high-class fish, so we, of course, had to fish the ponds. I never saw lads with rods bent like it! There were rudd and I think there was probably bream and other things, but certainly the rudd were quite big, a lot bigger than the ones we caught in the Lagan. But we only got that for a while, then they got wise to us so we had to call it quits up there!

All my mates were and still are fishing, we all still fish. The Lagan Canal, starting where Cutters Wharf pub is now – that was the first locks – that stretch of the canal up to the weir at Annadale, where the river goes out over the weir, there were barges there, and people lived on them. The canal was no longer in use, but they allowed the people to live on their barges until they got their day on them, and that river was absolutely stuffed with rudd. And we used to go up and fish for them, and a little float, a very small hook and a piece of dough was all you needed, and you would have caught them until your arm was sore. And when you got tired of that you could have got worms and went down and fished for eels in the first locks, some very big eels knocked about. But they always swallowed the hooks, it was an awful thing to try to get a hook back out of one. That section of the canal, it's now infilled, there was lots of barges there. To me, being a kid, there seemed to be quite a lot of them. They were all on the far side, not on the towpath side, they were all on the other side, where Rowan's house was, right up to the boathouse.

A little punt used to take you across to the boathouse. A little punt with a rope on it, a little thing it would have held maybe four people and if you didn't want to walk the long way over the first locks right up, past Rowan's house to the tennis club, you could have went over in the little punt, you pulled yourself over on the rope and there was a rope then went behind it, if anyone wanted to come back. We used to go across it just to say we had done it. Rowan's house would have been halfway between where Cutters Wharf is now and the tennis club. On the island, the far bank. The tennis club was a big wooden structure, it was burnt down as part of the Troubles.

There used to be hundreds of swans in the river, I mean in their hundreds, and they used to anchor up for the night, they all got together in a big, big group just at that stretch opposite Annadale School, that part of the river, the back of the tennis courts. They used to sleep on the water with their heads on their backs. And they used to sleep there at night. That used to be their gathering place. They just bobbed about. Always they were there. And now it's rare to see swans about. Always there were swans even upstream in the canal and that. Now there are just one or two.

Further up the towpath we didn't fish much until we got to the second locks and again, just above the lock gates that stretch running up, that canal was full of rudd there. And also, funny, Geordie Kilpatrick's, at the third locks, just at his lock gates, the stretch on up towards Newforge there, it was full of rudd. But Leister's Dam up at Bladon – the first water supply for Belfast – it was full of rudd and when we got tired of river fishing, we would have went up there to fish. And people didn't believe us that there was fish in it. It was a lot deeper and a lot bigger than it is now. I looked at it a couple of days ago, I was up with a friend of mine with the dogs and I says, 'You know, this is not the way I remembered it, it was an awful lot bigger than this and very deep'. The surrounding marshes, lower down where the black pipes are, below Leister's Dam, was full of hares. There were dozens and dozens of hares that lived in it. And every now and again guys used to go up – we would have met them when they were up – and they would have shot, well, in those days you just shot what you needed. If you wanted to eat a couple of hares you went up and shot them, now people would have shot everything, left it without anything in it. But any time you walked through it hares were getting up under your feet. Dozens of them were all over that area. Above those marshes where the black pipes are, up in the forests up there had been a golf course originally and you can still see where the old greens were.

The big house, further up towards Newforge, Gibson's house, where there is a golf driving range which is now for sale to build office blocks, it was always empty when I knew it. It was all there, it had windows and a roof, I don't really know the history behind it, what happened, but it sat empty for a long, long time. But Gibson's house was another place. I remember one particularly wet day we were up, all the kids were up, oh and it was lashing and we were just up along the river, and we decided to go up into the house. There were no doors on it and there were cattle inside it. We chased them out. So we decided we would play hide-and-seek and there was one wee character who lived in the prefabs at the bottom of Sunnyside Street, a wiry wee character, he was along with us. So, we were playing hide-and-seek but he disappeared. And then nobody could find him. We searched and everybody stopped doing what they were doing and we looked and we could not find him. And after about maybe 15 minutes we heard this noise. He had got into the fireplace and got up the chimney to hide. Well he was wet, he was soaking wet when he went in. But my God, you would want to have seen the cut of him when he came down covered in soot! We tried to clean him up as best we could, but we couldn't so we were going home anyway and we hoped the rain would have washed him, but rain does not do a lot for soot. So we got him to the bottom of Sunnyside Street, we got him close enough to the prefabs and we said, 'Right, you are on your own', and we sent him on. We left him to his fate and we all went home!

The Lagan Canal, to get back to the Lagan Canal, I only ever saw one barge actually moving on it, and I was very lucky to see it. One day, a motor launch was bringing a barge down and it was the last barge that was coming down to be moored with the others at Stranmillis and we saw it coming. And it was anchored up. But as the people died off or moved away, the barges then were done away with, they were broken up, but the very last one was moved – there was only one old boy left – and it was moved across to Annadale, and he lived in that barge. It was a big metal barge, a very big one. And he lived on it just opposite Annadale School. If you come out of the gates of Annadale School, there is a wall, the old wall is still there, a brick wall down by the river. That barge was lashed to it with big wire ropes and that. And we got to know the old boy – he wasn't a terribly friendly old character, but when he saw we weren't doing a big lot of harm up along the river, he sort of let us onto the barge to look at it. I remember it was all rivets and stuff like that. But then he died or moved on or whatever, but that barge lay there and rusted away and it broke in half and I remember the rusty bits of it left and I don't think there is anything of it now. Probably taken away and just scrapped. They have often talked about opening the canal, but sure, they had it open and they couldn't wait to close it![72]

The clay fields were as far back as that
Maurice Neill

My father was George Neill and his father was George Neill and there was Uncle Willie, Uncle Jack, Uncle Hughie. My dad George and they all worked in the Lagan Vale brickyard, so they did.[73] My grandfather and Uncle Jack and Uncle Hughie, they lived in 17 Wansbeck Street. My grandmother, she looked after Laganvale Band hall. I was brought up in 43 Laganvale Street, so I was. And it's still standing today, worth a fortune, like! But they all worked in the brickworks, the Lagan Vale, the one on Lockview Road. In those days if you wanted a job you just left and got another job, you know, if you left on Friday you had another on Monday. But I would have been, oh, say 18 when I started working there and was there until I was about 20 odd or 21. Around about that age.

The Lagan Vale brickworks. Behind the wagonway is the clubhouse of Malone Golf Club, which leased land from the brickworks. The photograph was taken before 1919, when the golf club moved to Malone. The clubhouse is now a private residence, at the corner of Richmond Park and Stranmillis Road.

You started work at half past eight and you worked to five thirty, you had half an hour for your lunch and on the Saturday you worked from eight thirty to twelve thirty. A man my own age, I asked him about what kind of money we were earning then. And he said we were earning about seven pounds per week, which was quite a lot of money if you work it out – 52 years ago. And they made the Lagan Vale 'rustique' brick world famous – they are the bricks with the wee dimples goes in them – and for every thousand that you put the dimples in, you got one shilling and one halfpenny, so you did. The dimples were on the outer skin. If you look at those houses or the walls you will find that they are, like, rippled. That was put on with a hand roller.

You started at half past eight in the morning when the horn blew and you didn't stop until half past twelve. You had a tea break, when I was having my tea someone was doing my work for me. The young boys, I would say 18, they were in the wet brick squad. They run the wet bricks until the drying kilns – I don't think that's the right name for it – they stacked them up on wooden racks and they dried. The older ones, they took the dry brick into the kilns where they fired them, so they did. And then older ones, well, they drew out the completed brick.

Cover of a product catalogue produced by the Lagan Vale Brick and Terra Cotta Works in 1909. Between the two statues is a sketch of the brickworks with the adjacent canal basin and Stranmillis lock.

Do you know the area at all? Well, Bladon Drive? The clay fields were as far back as that. They had a big mechanical digger and they had a little tractor engine on a railway line bringing it down to the brickyard. It was little – you know what I mean, what width is a tractor? It was a proper tractor, so it was. And they brought it down in that. I can remember as a boy going up to the brickfields and seeing the martins in the

Left: The old brickworks, Annadale, by Elizabeth Holmes Kinnaird. The painting dates from the 1920s or 1930s.

sandhills, but I think eventually they ran out of ground or ran out of clay. So that's what closed it. They also had men with spades to get clay out. They took it down to the wet brick area. What happened was there was a ramp up – now you are bringing back memories – and there was a little pulley thing that pulled the wagons up and they just dumped it into a big hopper. Hopper, is that the right word? And then it went down and then the next one came up. A machine pulled it up. Well, it was all churned up and came down and came through rollers and came out and then into a compressor. It was all compressed and then forced through this machine and it was all cut into bricks. They were put onto boards. When it came out on a slab it was shot forward: a chap pulled a handle, it went through wires – you know, like cutting cheese – and it went onto a board that was lifted onto a wee trolley. Now you are into the hand end: once it came out of the machine it was just big solid lumps of clay the width of a brick. It was forced through wires so the bricks were, you know, all the same size.

They were lifted by hand and onto a wheelbarrow, just flat boards. Then they were wheeled down 'til the drying area. It was just rows of sheds, wooden. And there were chaps actually lifted them off the boards, two at a time, and set them up, and then they were dried. They were steam dried so they were. They were there, oh, maybe a week. The dimples were put on while they were still wet.

Then once they were dried they were taken out and put into the kilns to be fired. They worked non-stop, night and day. They had to be. There was a man there to look after them. The kilns had big arches. I know that during the war, people went up, if there was a threat of an air raid they went up and into the kilns.

The boards I am talking about. Now, I don't have the correct pronunciation, what is it – I think people call it – creosote. Right? It's totally banned today. What they did was they painted the boards with that so the bricks didn't stick. You can imagine handling that! You got it on your trousers. When you went home, you didn't go into the house, you went into the back, the outside toilet, and took your trousers off! Because you could have stood them up on their own and the smell of it, like, I mean!

And they made chimney pots. They did all this fancy brickwork, you know. There was a man did it on his own. One man. And he was crippled with arthritis. His hands were. He used a big lump of clay. Set it on and hit the button, woosh. On a wheel. His hands were crippled with arthritis. That's one of the memories, you know, when you are a young man there are things you look at and you don't see but that was one I did. They used to send us round, say there was a slack period, they sent us round to give him a hand. While we would have made a quarter of one he would have ten made, so he would!

The kilns were massive, big. When they put all the dried brick in, stacked them all plus the chimney pots, there was a fire bed down each side but when they put all the brick in, they actually built brick up and closed the doors and covered it with clay. Then when the bricks were done they were not thrown out,

The last industrial building on Lockview Road, built for the soup works in 1950. It was later used by Lamont and is now offices. Photographed 2011.

they were set to one side so they just built the door up with the bricks over and over again. And there was a man looked after them all the time, so there was. One during the day, one at night. And the boiler man, he was there, it was covered 24 hours as they needed the steam to dry the bricks, so they did.

H and J Martin owned the other one, the brickworks, on the Annadale side of the river. They used to call it Bimbo Martin's! I don't know why. It sat on a hill. Haypark, I think that was the proper name for it, Haypark Brickworks. But everyone called it Bimbo's!

My grandmother looked after Laganvale Band, she kept the key and brushed the floor. My father and his brother were founding members of Laganvale Band. It was a flute band when they started, that was after the '14-'18 war. It's now a silver band, so it is. It's still there. They have won a few competitions in northern England.

The soup factory on Lockview Road, that was Edwards' Pickles, I am sure you ate those! Sure they were a big English company, Edwards' is still on the go.[74] They just made pickles. They called it the soup works, but I only remember pickles coming out of it. After Edwards' went, there was Lamont, the factory on the corner. The soup factory was further towards the river. They have kept the Lamont building standing, but it's now offices.[75] You know where the Cutters Wharf sits? See if you go up on the opposite side, where there are new flats and all, if you go up, there is a house sits out, on the right-hand side. And the Heavenly Twins used to live there. They were two old dears, so they were, and everyone called them the Heavenly Twins. No matter when you saw them they were in pinks and lovely blues. Two sisters.

In the lock-keeper's cottage at Stranmillis was a Mr Belshaw. The next lock up was the Rowans. My uncle and aunt. She was my father's sister. His name was Peter Rowan and she was Aunt Maggie. Then the next one up was Geordie Kilpatrick, he had a wee ice cream shop on the side of the wee cottage. There were two canal houses at Stranmillis, the lock-keeper's house and the weir keeper's house. Rowan later got the job as weir keeper. That weir is still there. They were semi-detached houses.

I remember the barges, so I do. You could have got on one. A decent fellow, he would have let you on, as long as you sat down, he would have taken it and let me off at my Uncle Peter's, you know, the next lock. And the lighter stable was where Queen's Rowing Club house is, that was the lighter stable there. They kept the horses, I would think there were seven or eight horses. There was kind of a pong! There was actually a public toilet with it, so there was. The trams stopped at the bottom of Laganvale Street and turned to go back. There was a green toilet, a metal green toilet. Anyone could use it but it was for the tram men. A men's standing-up place. That was one of the memories, because that's what you did after you had a swim in the Lagan. There was a water thing in it, you needed a key to turn the water tap on, it was for cleaning out the toilet, and the older boys they rigged it up. So when you had a swim in the Lagan you went in, turned it on and had a shower! So we were really up to date, so we were!

It was sold as Vulcanite
Patrick Grimes and John Johnston

Patrick. I started at Vulcanite in 1969 and it was in full swing. People brought in paper, newspapers and cardboard and old clothes, it went into the making of this fibre. It was fibre for making felt for roofs. The base of the felt was fibre and then they put the bitumen on it. Then the granules on top of that.[76]

Cardboard, clothes and paper. All second-hand, recycling. I don't know if they advertised for it, all I knew was it came in. In cars, vans and lorries, everything like that. It must have been word of mouth and people knew about it. They got paid by weight, they had a weighing bridge at the bottom of the yard, and you stopped and you got it weighed and you drove to the top of the yard and there was a big shed and it was piled in there. And they had people sorting it out.

There was men and women there to take the buttons off. They always said they got so much money in the pockets! That was a great thing! In them days, 40 years ago or more, they were getting all this money in the pockets! Stories of the people getting fleas from the clothes were true, God knows where the clothes were lying or where they were from, they were smelly. There was a fellow they called Tojo, that's all they called him. Hunter Collins was his real name, he worked in there all the time, he must be dead long ago. Carrie Bell, she was from just down the road, and the Carsons, they were from Lord Street in east Belfast; Mrs Carson, Norman Carson and Buttons, they called him Buttons Carson, that was his nickname – he was William Carson. The whole family of Carsons was there – the mother, the daughter and sons, there was four sons there at one time. Where they worked was a good bit from the main factory, there was a massive long conveyor, it was maybe 30 yards from the main factory, and the conveyor went right up into the building. It wasn't well-paid work in them days. I was always told in Vulcanite that you won't get a rise, but you can work long hours.

The rags were chopped into small pieces in a machine called a chopper, then it was sent up a conveyor belt to the loft and mixed with paper and water, broken down into pulp by a big wheel. The fibre mill had its own water supply from a tank on the roof. The tank was so big seagulls were swimming in it.

It went down to ground level where it was sieved to make it fairly liquid; anything not wanted was caught in the filter. It was pumped onto a lovely brass mesh and there was air sucking the water out and you could just see it going dry almost immediately. And then it became rigid. From being liquid you could actually see it sucking the water out of it and the fabrics gathered together. And then it started its way round massive steam-heated rollers and it started drying. And it became fibre. There was no glue. It was great to watch it, it was wet there and it was almost dry here because it sucked the water completely out of it. When it came out the other end it was rolled up into six-foot by three-foot rolls.

To make roofing felt, the rolls were taken over to the felt works. We put it onto a runner, a big axle was put through the centre, and it started to run down the factory. The new roll was joined to the previous roll with a hot iron and Copydex. It then went into a soaking tank then into rollers where a coat of liquid bitumen was added and green granules put on one side and sand on the other side. It came out very hot. Then the felt went over water-cooled rollers and was cut into ten-metre rolls.

The Belfast factory supplied the Republic and here, it was sold as Vulcanite. The bitumen plant worked 24 hours, though it closed Saturday midday until 6 a.m. Sunday. We also supplied liquid bitumen to a carpet tile company in Craigavon and to John Erskine of Whitehouse, a roofing felt company, and bitumen in bags was sold to Andersons in east Belfast. Then they made Perspex roof lights. They also made mastic, you put it on with a trowel. It was for fixing leaks. They made solar reflective paint. But the main part of their business was roofing felt, they could make 2,500 rolls a day of slater's felt, the mill could travel very fast on slater's felt that was backed with hessian and the light fibre.

The plain brick facade of the Vulcanite factory with the company delivery lorry, 1982.

Oh, the Vulcanite was a big business, they were really going well. It wasn't for the want of orders it closed down. There was a terrible lot of bomb scares in the '70s, but it was from inside, you were getting two hours, three hours off twice a week and they knew very well it was an inside job. Someone who was on nights might phone up and get them all out! It was a combination of lots of things. It needed a lot of money spent on it, it needed millions spent on it, it was very old. It wasn't a big site, I would say around about three acres. We used to have explosions, you see, we had massive big columns, columns about 40-feet, 50-feet high, and that's where we processed the bitumen, you blew compressed air through them. Say, if fumes were not getting away properly or if they were blocked or whatever, the lid on the top blew and you got showered with bitumen! Cars and everything at the adjacent houses because it was so high up, at least 40-foot high, maybe more. Two columns – they called them columns but they were old-fashioned boilers turned on their end. We had claims, yes, for cleaning cars. There were also complaints about smells from the afterburner when it was not working properly, and there was a terrible smell from a heated tank where we mixed rubber and bitumen to make a special kind of felt.

Around 1974, part of the factory, the fibre mill, closed. From then on the fibre was brought in. They sold the fibre mill for scrap as the drying cylinders were made from gunmetal and were very valuable. We were told the roofing works was closing about April 1989, though the factory went on working for some time to use up raw material. I have a little memento of the building when it was pulled down. I have a piece of wood with a man's name handwritten on it, Joseph Fitzsimmons Junior, a man who I suppose helped build it, and the date, January 1900.

John. There hadn't really been an awful lot of input into it, the building, the facilities. The wee road and all to the Vulcanite factory, right at the end of Lockview Road, was all lumps and bumps, pieces took out of it which needed filled in. The factory was in very, very bad condition. I started in 1986. The main thing they produced was roofing felts for under slates and torch-on felts. There was a chemist there called Andy Thompson and he pioneered a new material, it was actually to do with the chemistry of polymers. He worked out this method, this type of material which was used with one of the felts. It is still used to this day. It was a new type of surface on roofing felt. They also made aluminium paint, primer, felt adhesive, mastic and a few other bits and pieces like that.

Whenever I went for the interview, I had a very old map of Belfast which I took out to see where it was, because I didn't know where Lockview Road was – I live in Carrickfergus – and when I looked at the old map it actually had marked that area as a paper mill, which obviously had been there at some stage. When I went along, the paper mill was gone. But I was told stories from older men who were in the place who had actually worked in the paper mill, it made fibre and it had been done with rags which were supposed to be leaping with fleas, you know! The fibre was used as base for some of the felts. Still to this day there are fibre-based felts, the type of felt that you would get on a shed or a dog kennel, nail-on felt with mineral on top of it, now that would actually be fibre based, but that fibre can tear.

There were 55 people working at the Vulcanite when I started there. They talked about the shop. It wasn't a shop, it was actually a workplace. It was very, very antiquated, it was just basically brick walls with old-fashioned type windows which had been replaced with a plastic type of thing that Vulcanite had actually made. It was reinforced plastic – not plastic, it was made from one of the poly products – you know, it was like translucent with reinforced diamond mesh inside it. And inside that building it wasn't even level ground, it was sort of up and down, it was a long and narrow passage, and on the left was like a roller system where the felt was manufactured. The fibre polyester went up round these rollers, in through different various tanks with bitumen-based products and then came out at the other end. It was a single-storey building, very tall with extractors up at the top to extract the talc, the chalk – that was part of the process, too. In summertime you could have seen the chalk and the dust in the air with the light coming through the windows. Rather similar to an old picture house years ago where people smoked and you could see the smoke with the light from the projector, that's how I would describe it to you, you know?

The residents at the back in Sharman Road, now I know they complained about the smell. That was an ongoing thing. And actually in dark nights when I was driving the forklift truck – there was like a perimeter road down there – big, massive stones would have bounced off my forklift truck, they would have injured me if I had been hit, and I knew they were coming from the Sharman Road direction. I knew there were young ones up there, sort of standing looking over the wall. In the area behind the shop were tanks of crude oil and bitumen, big vertical grey metal tanks, and a big metal hopper that was the sand and granule separator. They were not visible from Lockview Road and most people did not know they were there.

The factory was run down. When I first started at Vulcanite, the forklift truck was a terrible thing. For one job I had to go up a ramp and lift a barrel of oil up onto a shelf, it was about 14 feet high and the clutch and handbrake weren't good. The ramp was only about the width of the forklift. It was really dangerous.

Fuel oil and oxidised bitumen tanks at Vulcanite sketched by John Johnston, 1988.

The factory closed completely in 1989, there was a very bad accident in the factory, a very bad accident indeed. Production moved to England, we are now just a distribution centre at a warehouse in the docks, it is now called ICOPAL. But one thing I did when Vulcanite closed was I took two of the cats; there were a lot of feral cats and I brought two of them to the new plant.

My grandparents lived in the second lock house
Jean McDonnell

The lock-keeper's house and weir at Belvoir Park, close to the second lock.

I was born in the caretaker's cottage in Belvoir Park and grew up there with my parents and sister Joan. The cottage was in Ballylenaghan townland, on the Newtownbreda to Purdysburn road and past the main entrance to Belvoir going towards Purdysburn. It was behind the estate wall and was not visible from the road, there was a small gate in the wall and near to it big, red wooden gates. A small stone house with small little diamond panes of glass, one storey, slated. It had a large living room and two bedrooms, gas lighting and no running water, we got water from a spring well. There was a large garden with a shed.

Robert and Elisabeth McCurley.

My father, Charles Darling, was a plumber but had the house free because he looked after the place; he would see if there were any trespassers and would open the gates for to let cattle into the different fields. You see all the fields were let out. This was in the late 1920s, when the big house was empty. The estate belonged to Lord Deramore and later Stewart and Partners. I remember the entrance gates, I suppose the people that lived in the wee houses at the entrance, the ink pots, also worked in the estate.

I was all round the grounds at Belvoir. There were the ruins of an old house at the top of the orchard, did you notice that? By a wall. There were the remains of an old house in the woods in the south of the estate that I think was called Mr Byers' house. There were ruins there and I remember my mother saying that was Mr Byers' house. I was in the big house but it was just empty. To me, the house was like a dungeon. I was in the outhouses as well. I remember seeing elephant's feet, antlers and things like that brought from abroad lying about. We used to be fascinated by them! I don't know what they had been for, something left from years gone by, you know. There was nothing happening in the house. All was quiet there.

I remember the old orchard very well. I lived on the fruit. There was every sort of fruit you could mention. Everything. It was overgrown. There were lots of apple trees, pear trees, plum trees, you name it. I always remember the kemp apples. They have sort of vanished off the market, they have a different taste, different texture. At the very top of the orchard there was a big cherry tree and redcurrants, blackcurrants, gooseberries. There were old greenhouses in the gardens at Belvoir, but they were all tumbled down.[77]

I remember the old graveyard well. The gravestones were standing then. My father used to come that way often because he rented a field. Do you know where the sewage works is? Now he rented that area and he would have come that way quite a lot, you know, when he was going home. And this night all these lads from the village were coming, heading to the graveyard, you see. So, dad had a white shirt on and he took the jacket off and he got up on the wall – well, it was rumoured from the one end of the village to the other that there was a ghost in the graveyard! The graveyard wasn't the way it is now, it would make you weep.

A man who lived in Newtownbreda Village, a man called Fred Guy, looked after letting out the land at Belvoir, my father was just a caretaker, he stopped trespassers coming in and such. Mr Chesney was also employed at Belvoir, I think his parents lived in the ink pot gate houses. There was a man, Sullivan, who rented land towards the river, near where there used to be the little sawmill. The land between the canal and the River Lagan was called the Island, they had to swim the animals across to it, there was no other way of getting them on.

People that shipped a lot of cattle would have drove cattle to Belvoir for two nights or a week and then away again. They must have come from market or were being shipped out. There were no shoots at Belvoir when I was a child, all that had stopped, but I can remember seeing a fox hunt at Belvoir, just once. As time went on, all these kind of things fizzled out.

My grandparents lived in the second lock house, their names were Robert and Elisabeth McCurley, they had two girls and two boys. He opened and closed the locks. The cottage was big and roomy, it was upstairs and downstairs. It was divided into two or three rooms upstairs, but there was one enormous big room downstairs. And then that little bit, it was like an outhouse, they kept that as a wash house. They had the land at the house as a garden, and granny had a vegetable garden and fruit, lovely gooseberries, we used to steal them! She kept chickens as well. They weren't very rich, but my grandmother was a very well doing sort of a person, she baked and made her own jam and

Elisabeth McCurley.

done everything. My grandfather opened the locks which were a good bit down from the cottage and he had a little hut down beside the lock, that was where he used to go for a sleep! He sat there, you know, and watched for the barges coming up. Not hard work, good sleeping time! But she made him toe the line, she was boss. Definitely. They left around the early 1940s, I would reckon, though I can't be one hundred percent on that. Then the man Rowan looked after the locks, but I don't think anyone lived in the lock house. My grandfather's uncle, Taylor, worked there before my grandfather, and that's why it was called Taylor's Lock.[78] He and his wife, the Taylors, were the last people buried in the Belvoir graveyard, in the 1920s, I think. I think there was a headstone but the graveyard is all overgrown now. The graveyard was private, it belonged to the estate.

The monk's well. How can I describe where it was? It would be to the right-hand side to the orchard. Down the slope from the graveyard. There is a small stream in the area and a little bridge and a wood to your right, and I think the well was in through the wood near there. It was called the monk's well, the monk's wishing well. I think my mother told me someone came looking for it some time or another, so it must have been known about, wherever it is, in the woods.[79]

I used to watch the people, they called them inmates in them days, in Purdysburn Asylum working in the fields not far beyond our house, they would have been gathering potatoes and things like that. The keepers and all were with them. It is changed days now. They were like prisoners. There was a loud bell they would ring to let you know if somebody was on the loose. They gave the bell a blast and that was it. We knew to go home. They may not have been dangerous. There was one person who lived in the woods for a week before they got him. Nobody ever bothered us. You wouldn't have the fear of them nowadays, it's the ones with all of their senses you watch out for now!

We went to Newtownbreda School, we had to walk it. It was a very happy childhood, I have very vivid memories of being happy there. In fact, the most vivid memories of my whole life. Play in the woods. We went out in the woods and gathered blackberries, raspberries and things like that. Like it was a lovely childhood, it really was. At the top of the garden we had big beech trees, at night-time the badgers would come out, you could have watched them for hours. Things like that. We gathered fruit, flowers, bluebells, primroses. We would have cut right through the estate to visit my grandmother's.

We were still in the house at the start of the war. The estate was all wired off, you see, there was ammunition and all stored in it. The soldiers used to come to the house for a cup of tea. Barbed wire all around the place. English soldiers, I think the first lot that came, I think they were the Royal Engineers. My mother would make them tea and give it to them through the wire. Some of them were just boys. We left the house in 1942 or 1943, we went away out beyond Ligoniel. No one lived in the cottage after that, it just fell down.

Byers was the gamekeeper when we were at school
Albert Allen

I first visited Belvoir Park when that one – remember Stewarts the builders – was going to build on it. That's the first time ever I remember being in it. For when I was young, you know, you daren't look over the wall. There was a keeper there, they called him Billy Byers. He was very strict about looking over the wall. And the man who was in the second locks, Bob McCurley, he was nearly as bad! He wouldn't have let you over the second lock. But anyway, Billy, I think he was a very nasty sort of a man. I have heard that from one or two. Geordie Kilpatrick the lock-keeper at the third locks, he had crossed him two or three times. Geordie said, he told me that he was standing at the door this day, and five or six pheasants flew past him and lit in an old boggy patch up from his house. And Geordie went in to get the gun to have a shot at them, and Byers come on the scene. Well, Byers kicked up a whole row about Geordie and Geordie maintained – well, when they weren't in Belvoir he was entitled to shoot them, you see. But anyway, Byers wasn't pleased with Geordie, anyway. And the next thing was, Byers went to the land steward about it, about this, and they called the land steward Fred Guy, he lived in Newtownbreda. He come and seen Geordie. I think Byers was wanting Geordie put out of the house. I think Guy was a wee bit sympathetic towards him, you know, and he told Geordie like not to let it happen again. But Geordie and him, Byers, never agreed.

Mr Byers the gamekeeper.

George Kilpatrick, keeper at the third lock.

Then there was another fella, you called him Ben Cardwell, he was a great big fella about six feet, six feet three, and he never worked, for he done an awful lot of poaching in Belvoir. So Geordie didn't want him coming about his house too much in case Byers traced him down into his house. So anyway, Ben Cardwell, he went in as often as he could get the chance. And this day he was in and there was people who owned the land around there, they called them Carsons, they were butchers. Well anyway, Carsons had a big dog, and I believe he wasn't too good of a tempered dog. And he was caught in one of the traps the day Ben Cardwell went in to do a bit of poaching. And Ben let him out of a trap. And anyway, when Byers come on the scene, Ben made off and Byers was in round by Geordie's house, I think there were words then too, you know, about it. Geordie never liked him.

Byers was the gamekeeper when we were at school, when I was about 14. I was born in 1911. Well then one time there, down at one of them streets, there was a family come to live, they called them Brown. And there was two girls and two boys and the father and mother. Well the father was very, very old. But a small man. He wore an ol' hard hat, but as green as grass. He was a keeper in Belvoir – no, not a keeper, a shepherd. A shepherd in Belvoir. And he was in it for a lifetime. And he was well up in the

Bible. And he used to tell us when he was minding his sheep in Belvoir, he used to read a lot of the Bible. And I think one of his sons is still living, the Jim one. I was at school at the time, I was about 14 and we used to speak to this man Brown.[80]

Well, you see, I was in round by Belvoir during the war. I went in with a fella, he was going to shoot rabbits. I wasn't a bit keen on going in, but I went in with him anyway. He was a fella lived on the Lisburn Road. And the big house was very derelict. And ol' bits of trees and all was growing up against the front door. It was terrible neglected, you know. It was in disrepair. We didn't stay long in it for I didn't fancy going in it. Shortly after that the army moved into it and I think they ruined it completely, you know. And then they knocked it down. But there was one of the girls that was at that exhibition that I went to see in the library – she works in the library part-time – and she told me that it should never have been knocked down, they were far too quick in knocking it down. She was all on for preserving it. And for she said it had a great history, you know. I met a man in it, too, one time. And he said to me, he said he was 70 years of age and he said his mother was a milkmaid in it. I would have liked to meet him more, but I never met him again. I would have liked to have found out a wee bit more about it.

Geordie Kilpatrick told me that that one Billy Byers lived in Belvoir a short distance from him, up on the hill in Belvoir, among the trees, and he said the ol' house he was living in was a wooden structure, and you wouldn't have wanted to go, it was a very derelict old place. Then they moved in, I think, to the house that's still standing at the car park, they moved in there. They said there was two of his daughters worked in some cake shop down the Ormeau Road. Somebody told me that, but whether there is any truth in it or not, I don't know, you know. That was a good while ago.

I remember the old sawmill, but it wasn't in working order then. They had a kind of a wee river run in from the main river at the new bridge there and it run in and worked the sawmill. And Geordie Kilpatrick told me that there was two brothers used to work it in his time, they called them Wishart. Did you hear of them? The wood from the mill was used for fencing and also for wood for the big house.[81]

But Geordie had a more colourful life. I remember Geordie telling me he arrived up at the weir this time, and there was a pram coming down towards the weir with a baby in it, a wee baby. And he got what they call a big fork, it was an awful length, about eight or ten foot long with a hook on it, and he got the pram and baby out. And he looked up the river at Shaw's Bridge. The mother was in the water, but he couldn't get her out. And he done his best to get her out, but she was drowned. And it appeared that she had had a row. She lived in Balmoral Avenue and she had a row and she went down to drown both her and the baby. And he told me that at the inquest they said about them having a row and Geordie said that the husband never even came over to thank him for saving the baby.

Bob McCurley lived in the second lock at that time. His wife Elisabeth was a Scottish woman and she was a very nice person, a lovely person, but he was gruff, he was very gruff and all. He wouldn't have let you across the locks for any reason. But Geordie Kilpatrick told me one time that he went up to the weir and there were about eight or nine big shoeboxes, cardboard boxes, floating down the river. And he fished them all out and he sent for the police. And they had been stolen out of a warehouse in the centre of Belfast and when they opened them up, the whole nine boxes, they were only for the left foot, there were no right-footed ones at all in them! All left-footed shoes and boots, and there wasn't a right-footed one! So, the police said whoever stole them, they were no use to them.

But I would have rather had Geordie than any of the other lock-keepers. See, it was a big family, you know. Six girls and four boys. And I remember walking up past them with a friend I knew, and he said, 'Look, they are already like steps and stairs'. Geordie said to me one time there was a woman on the path all morning and he seen her passing the lock house quite often that morning. See in the afternoon he was pulling her out. He was never done pulling them out, Geordie. But Mrs Kilpatrick told me he was very hard with the money, he wouldn't have parted with it at all! And ol' Bob McCurley that was in the second locks, that's the one at Belvoir – he was a gruff old fella. I didn't like him at all!

The last one on the second lock house was Peter Rowan. Well, old Peter, he was always moaning about the family. He had four of a family. Two boys and two girls. And I knew his wife, she was a very decent sort, Mrs Rowan. Well then, he complained and complained about the isolation of that lock house and that they couldn't get to school. And then they put him in a house there where the car park is there in Lockview Road. There was three or four houses there, they belonged to the Lagan Navigation and anyway, they put him into one of them and he brought an ol' kind of a diesel engine to cut blocks up in the house at the second locks and he cut blocks and all. And he had old diesel oil in it and here – one time he went down for his dinner, and the old diesel oil went on fire and it was burned down. The lock house was burned down, burned completely. Destroyed every bit of it. Somebody told me they were passing at the time, and the slates was flying everywhere. It was burned completely down. He was the last one in it. Bob McCurley was there before him and he had looked after it.

It burned down sometime after he came back from the war. He wanted me, you know, to join up. And I said not at all. And we lived in Dunmurry at the time and he wanted me – this is before the war in the beginning of 1939 – he wanted me to join up, in the Territorials or Home Guard. And ach, I put him off, anyway, I didn't want to join up. But anyway, after the war started and all, his brother-in-law went into the job, in the locks. His brother-in-law Jack Neill. And I said to Jack after the war started up a few months, says I, 'Where is Peter?' He said, 'He is away in France in the thick of it!' That was the Home Guard!

Peter went to the war and whenever he came home after the war was over, there was a fellow from the South of Ireland at the lock. And he was doing a good job at it and was attending to the work in the second lock for the barges, you know. Billy Hughes. Billy Hughes says to me he was talking to a man, Mr Patterson. And he says, 'If that fellow Rowan comes back, you'll be put out of the house.' So Billy went down and seen them in the navigation company. They said, 'You'll not be put out at all when Rowan comes back. We'll offer him another job and he'll just have to take it.' But anyway, Rowan came back and Billy Hughes wouldn't get out. And Rowan wanted into the house, and then there was that thing after the war, reinstatement in your job, you see. He took the Lagan Navigation to this reinstatement tribunal and he did win the case, and Billy Hughes had to get out. And they sent him away up to a lock that was vacant a way up the high bridge near Lambeg. That's where they sent Billy. Then, I think, with ill health he went down to live in the South of Ireland again. Old Peter Rowan. And a whole lot of times the Lagan Navigation wanted to sack him for he let the ol' weir flood, he didn't open the gates and he let the weir flood.

And there was one time – what year was that, about 1937? There was a terrible wet day, in the wintertime, and Rowan went away down to town, he spent all day in the town, and whenever he came back his wife and her brother and all the youngsters was up the stairs, there was three foot of water in the kitchen. Did you hear about that? She was isolated for a week in it, they could hardly get food over to her. They got a rope over and then they sent her food, got food over. And they were isolated up the stairs for about a week before they got them. And he couldn't get over to them. And his mother-in-law cursed him into the hot place and out again! After about a week they got them out, you know?[82]

But I remember sometimes when we were round the towpath we would see a few pheasants. I have seen the odd hare and I seen the odd mink, plenty of foxes and plenty of badgers. But the time Byers was there, sure he shot everything. Waterhen and coot. Everything he shot. It wasn't right. Do you remember in Belvoir, the rabbit warren? Oh, and the rabbits were in hundreds in it. As you go over that new bridge and turn right, you know the new bridge, the wooden bridge at the second lock, if you go over that from Stranmillis, turn right and then turn left, and away up on the hill. There was millions of rabbits in it. I never seen as many rabbits and one morning my brother and I was up there very early and the whole field just moved. Ach, it was sad about the Myxomatosis, wasn't it? Ruining the rabbits.

The front door was never closed through the day
Dorothy McBride

Well, I am Dorothy McBride, I was Dorothy Kilpatrick. Do you want my age? I am aged 69. I was born in the third lock house, ten of us were born in that house, in the same bed, in the same bedroom. And all reared there. My father was a lock-keeper.[83] He had a wee shop, he had pigs, he had a garden because the wages weren't very high then. So he had to raise his own vegetables and he had two goats for milk, which I hated, I hated the milk! He worked in the garden between barges coming through because there wasn't money coming in and the house was tied to the Lagan Navigation Company.

George Kilpatrick in his shop.

My father's name was George Kilpatrick and he started as a lock-keeper around 1921 or 1922. He was there a short time and then he married my mother, who actually worked in Lisburn as a cook. They married in St Anne's Cathedral, though it wasn't a grand marriage, and the honeymoon was to walk down to the lock house to live! No honeymoon or anything like that! My mother was Sarah Clarke before

Three of the Kilpatrick children on the lock by their home. Georgie aged about 18, Dorothy about 9 and Margaret about 6.

she married my father. There were ten children. The eldest one was Nancy, the next one was Ruby, the next one was Winifred, the next one was Georgina and then there was Hugh, Ernest, Jim, Stanley, Dorothy and Margaret. The older ones, you know, sort of looked after the younger ones. My mother had her own geese and she had turkeys raised to make a bit of money, that was her wee bit of money at Christmas. But we had no amenities at all, there was no electric. My mother cooked on the fire. It was an open fire. Big, big open fire then. And my father would have been allowed coal off the boats, that was one of his perks. Massive big coal and he would have cut them up. They were just thrown off, lumps of coal. And then he cut logs. There was two hobs, one either side of the fire. The kettle was always on one, teapot on the other. There was a big fender at the front and that's where she kept the plates warm. She had a griddle, she baked every morning.

Before we were out of our beds she had baked soda bread, potato bread, wheaten bread. How she ever done it, I never know. We didn't think about it, it was what she done. My father liked the bread. We come up to the shop, we were sent up and would have got a plain loaf it was called in those days but it was mostly home-baked bread we had.

And she would have done bannocks – a Scottish thing – wheaten and flour, butter and margarine and stuff like that. And put it on the griddle. The griddle was always handy. Potato apple, you know potato bread baked with apple in it, and that was a treat. My father had all the bushes – raspberries, gooseberries, redcurrants, blackcurrants. Strawberries. And he grew his own peas, rhubarb, grew his own potatoes, cabbage, lettuce. And the potatoes were stored in what you called a bung in those days. We just called it a bung, I don't know why. It was soil, a mound, and inside there would be newspapers. And there were different names for different things. There was like a dump, it was called a dunkle. The things that we say, it was just the way we picked it up as you go around. We were considered country people then. Now it's close to Belfast. You weren't considered close to Belfast when I was being brought up, it was like another country. You only went there when you had to go, when somebody took you there.

My father was a bit of a character, you see. The old type, and people liked to listen to him. He always walked with his hands behind his back, both hands behind his back, walked up the towpath, you would have seen him coming with hands behind his back and with a cap. He had a good hat if he was going anywhere, a paddy hat if he was going anywhere, but he mostly wore a cap, I used to buy them for him in Gordons in Sandy Row. I remember going for them. My mother was a small stout lady – having ten children you could imagine – and she always wore a jumper or blouse with an over the head overall, a dark overall that you tied at the back. She didn't wear dresses nor anything like that. That was what she wore all her life. Very old fashioned, very dark in colour. She had dark hair, straight back, no curls then. A very nice lady. I don't know where she got her education, but she was quite clever. My father was very clever at counting, he could have counted anything. But reading – she read the paper to him. She read the paper or anything that came, she would have read the invoices for the lemonade companies. My father always dealt with the money, he knew what to do with his money! She taught him to sign his name, I remember her teaching him to sign his name. But a clever man in many other ways, he just never got the schooling, he was working from when he was no age. We thought nothing of it, but he did make sure we went to school.

Sarah Kilpatrick.

There was the shop, like a wee shop at the side of the lock house, it had a sloping corrugated roof. My father ran the shop, another way of making a bit of money. He sold lemonade, cigarettes, sweets – big bottles of sweets, chocolate bars. Newspapers at the weekend. All at quite reasonable prices because people weren't well off. There used to be seats outside and they would have sat, and my father sat among them. If asked they would do cups of tea, but my parents weren't keen on that because there was that much to do about the place. It was mostly cold drinks. But I remember the time Coca-Cola came and he got this big container like a big freezer, but you had to go to the ice works to get ice for the weekend. And he put the Coca-Cola in the ice because we had no electric to have a fridge. I was not very old at the time, but I remember it happening, someone taking him to the ice works around Shaftesbury Square. The shop was open every day, but Saturday and Sunday were big days. At that time people walked more than they do now. People always walked because that was the only way to get about! And the courting couples, like, you know. Myself and my husband too, we walked those towpaths.

And up the lane, going over the wee bridge, would have been the wee cottage that used to be, I am told, the lock-keeper's cottage a long, long time ago. That was going up towards Milltown, before you came to Spence's, it was the first wee cottage. My sister lived in it for a while after she married, but there was, oh, a succession of people in it before that. Someone told me it actually was the original lock-keeper's, before our house was built, but my father never lived in it. It's been demolished now. The whole lot was flattened. The lane from the lock to Milltown was called Kilpatrick's Lane when we were young, but its proper name was Logwood Lane. The Spence's was a lovely long cottage halfway down the lane and they had a lovely garden. Three sisters. One of the sisters had a child, Marion. They lived there. But it was a lovely cottage. I used to go in, we were sent in because one of the sisters was crippled and we always called – going to school, coming home from school, in case she needed anything. You know, that neighbourly thing, you just did it, that was it. Their names were Sarah, Lizzie and Maggie.

Molly Ward's, the first lock, I didn't know much about. My father talked about it. But – what did you call the lock-keeper at the second lock – I knew him very well. Rowan. My father was on call 24 hours a day and if the river would have flooded, the weir had to be pulled as they called it, these sluices had to be pulled. If Mr Rowan was not available father had to cycle to his lock and pull the sluices for him. Now, how they got word to each other, I do not know because there were no phones.

The lighters would have pulled over at night above the lock sometimes, stayed overnight and travelled on. It wasn't fair to the horses for to travel through the night. They usually pulled in when it got dark and, if they were at our house, they would have come in and had tea or got tea and all in their cabin. They had cabins in these boats, they slept on the boats. Wooden bunk things, they weren't – certainly weren't – very dressy, but they were functional. I remember some of their names like Ned Larkin was one of them and Black Jack or Black John or something he was called because he had a beard, a long beard. And I remember them because they would have come to yarn to my father, sat in the shop with a lamp, just a storm lantern and yarn until all hours of the night. A lot of it was through the day and it was coal. Coal and grain taken to somewhere in Aghalee and, I think, wood sometimes. Sometimes grain, I always remember it because my father would have been sort of maybe looking for a bit of coal, and that boat wouldn't come through 'til the next day, you know. But how they knew, I don't know how they knew, because

The weir where the canal and the Lagan met, about halfway between the third lock and Shaw's Bridge, in the mid-1950s. The weir was managed by George Kilpatrick until it was removed in the 1960s.

there weren't phones. So someone had to pass messages along the line, as they called it. You know, I never knew, I never thought about it. I can remember there used to be about five boats a day in my time, probably more before that because it had started to, you know, sort of ease off, there was more transport on the roads then. It did ease off a lot and then there would have been maybe only two or three, you know. Then everything closed down. Quicker, cheaper, probably, to go by road.

A bit further upstream, where there were houses where people called McBride and Brown and Jack Daisy lived, you went to the back of those houses round to the weir. My father looked after that, it was his weir. It was almost up at Shaw's Bridge. There was a footbridge across to that, we were never allowed to go across, it was dangerous. There was like a wee race they called the water that was run off it. There was a whirlpool at the weir, it was quite dangerous and we weren't allowed to go there. We did sometimes but we weren't allowed to! I remember it flooded twice, but a lot more before that, my older sisters would remember it better.

In winter my father had to be almost never in bed, hardly never in bad weather. He had a thing, a key they called it, for pushing the weir up and down to let water through and things like that. If it was raining and the water started to rise, he was up at the weir. He had a horrible job, soaked to the skin manys a time. My mother used to be up with him too because women then supported their husbands! She would have been up having tea and stuff ready for him. The lights were never out hardly when it was bad weather because he would have travelled to that weir. As he called, 'I am up to pull a couple of sluices.' And that was to regulate the water. Until that new weir come in it was always flooding. Right up into the house. I remember it being very high, I remember being carried through and up the lane. Once you were up a wee bit you were out of it. Twice I remember, but I am sure there were a lot more times.[84]

There was a boy who lived in that house where my sister had lived, in the lane, who had been playing with a wheel and a bit of wire and he was only four and he drowned, in at the lock. It was a wet Saturday, I can remember that, and nobody missed him. My father pulled him out, he had a big long hook to pull him out. He was one of the Gribbins. Four or five years old. My sister and myself had been to the pictures on the Lisburn Road that day, when we came back this had all happened, you know. My father was always pulling people out. As I say, I never saw any of them because as soon as anything – my mother then took you in out of the way, you weren't allowed to get involved in those things. At different times people, a body, would come down the Lagan. Then there was a young fellow at the weirs, he got in, swimming, and it was the whirlpool he was dragged into. Now, he lived up at what they called Taughmonagh, up there the upper Malone. They used to come down there to swim. Again, I remember it, so I must have been about 12, 13. He was drowned. Somebody raised the alarm, somebody with him, I take it. Father went, again, and I do remember actually running up after him and being sent home. I can remember that very well.

The lock-keeper's cottage. Well, you walked in the front door and in at the back of the door would have been a storage place – not a big storage place – like a cupboard, doors and a couple of shelves. And then there was what they called the hall door, a wee door with glass – because the front door was never closed through the day. There was a chair always sat just beyond it, and my mother when she got older used to sit and look out at anyone coming or going. The wee window gave a bit of light because it was very dark. And under that as you went in, sitting at the side of the fire was a bench where you stored – it was one you lifted up and put blocks of wood and stuff like that inside. And you sat there. That was one of the seats beside the fire. Beside the fire would have been another storage place where they kept the shoe polish and all kinds of stuff like that. A table at the window, this big massive fire, a sofa and a chair and a table. There was really no room for anything else. Then into the room where my mother and father slept. The stairs were up against the wall, they were just like a ladder closed in, put up through the day and taken down at night. And there were two attic rooms up there, boys in one, girls in the other, two double beds in each. We thought nothing of it. In each gable was a small window. It was all oil lamps –

Right: After decades of neglect, restoration of the third lock started in 2009 and was completed in 2010.

there was no electric – and candles. My mother and father were in that room and I don't know what age we were when we left her because there was no cradle or cots or anything in those days, you know. You were reared practically from when you were born in that bed until you were old enough to move on. It was primitive in ways, but we lived, it was very happy. Hard-working decent people. Up early at six o'clock in the morning to take my father his breakfast and away and back in by the time we were out of bed. And she would have had herself all washed and dressed. You never saw her sitting about, there were no dressing gowns or anything then.

A tin bath, a tin bath in front of the fire, it had to be when the boys were not about, you know. When you got older you took a smaller bath up the stairs with the kettle. And yet we always managed to be all right. It didn't do us any harm. Mind you it was a chore, but you didn't think at the time. I think there was one mirror in the house, everybody fought over it! I think the boys were the worst!

My father had cattle, a couple of cows at that time. He only had the bit to the bridge for a garden, but he would have leased a piece towards Minnowburn there, you know, at Minnowburn River, he would have rented that from someone. I don't know who owned it. Later on it was just bullocks he had, not milking cows, bullocks for a bit of money. And then he would have cut the grass there and somebody, the like of my brother-in-law, would have came with the ruck shifter to move the bit of hay, you know, for the livestock back home. Because there was pigs, he always had pigs down at the side of the house – not the side of the shop but the other end of the house. There was what, one, two, three sheds there and the outside toilet. It was there, too! He had these pigs and mostly one cow at a time and when they stopped milking he would have changed it. One cow at home, the rest at Minnowburn.

My mother had two or three geese and she always had turkeys and hens. Five or six turkeys for Christmas. And the geese were horrible things! They would have attacked you! She only had two or three of them. But an awful lot of hens for the eggs, again for the family. Oh, I used to hear them talking about the foxes, they used to come up and raid the hen houses. I remember that, now, and the garden, we had to help to do the weeding. Of course we pinched a lot more, you know the peas, fresh peas, we used to get our knuckles rapped for eating them.

The canal shut finally in the '50s I think it was. My parents were both still alive then. My father died first. He still would have looked after the riverbank and kept the weeds down. It had shut but the water was still there. The Ministry of Agriculture took over. He was paid a small amount of money for that. It wasn't a lot of money, only keeping the water level and keeping the place scythed, along the edges, keeping it tidy. My father was devastated, he didn't think it would ever happen, you know, it was his life. He never knew any other life, his whole family were involved in the Lagan, you know, they lived in Aghalee and they were all lock-keepers and what you call rivermen working on the repairs and things like that. So it was a big shock. The lock gates were kept on and he looked after them until they took the gates away, which they should never have done, I remember dad being annoyed about that. He was still there when that happened.

We used to go up as far as Edenderry, we were allowed to go as far as Edenderry on a barge as long as there was three of us. You were put off at Edenderry and we would come back down, usually we had wee old bicycles, old rackety things. That was when you were getting on nine, ten. You were allowed to go. The rivermen were lovely men, they looked after you, made sure you were all right. Only one that I remember had his wife with him. My God, she was not like a woman! A very tough lady, she travelled with her husband, she wore dark, long clothes and all, and I was young then. I can't remember her name but she travelled with him. But I do remember a woman, she was so suntanned that she was nearly mahogany coloured, because she was out in the open. They had a wee stove and all in their cabins, people survived with very little in those days. They seemed to manage. I don't remember children on the boats. They would nearly always have dogs, they would have got off at the locks and run around and all that sort of thing. They nearly always had a dog. The horses all had names, Nellie and stuff like that. They were beautiful, nearly like shire horses, you know, big hooves, like, lovely horses, very docile. Now, they had to be rested and hay given to them and things. There was the odd time there was two, depending on what they were pulling. Steel, if they were taking steel they would have two. But mostly one horse. I think the boatmen were employees, I don't think they owned the barges, the lighters. Always the same man with the same boat and they would have named the boat or barges – they were called lighters, actually, lighters. They always had a name and it was mostly women's names. Always the same ones on the same boats and they worked for years. The horses were very precious to them. There was one who walked with the horse. There was sometimes three would have been on the boat, but a lot of them would have got off and walked the horse for quite a bit, along the towpath. I remember that so well. We would have run down when my father said there was a lighter coming, my sister and I and maybe my brother would have run down halfway, round the corner next to Stranmillis, and would have found them walking with the horse there. But they were very, very good to their horses because that was part of their lives. They were nice horses too, working horses.

After the canal had shut, all of the children had left except one, Stanley, who didn't marry. But then what you called the doggy men, people who had greyhounds, came and asked could they use what was the shop to keep dogs. The greyhound men also had goats in the field by the house. It had got very, very lonely down there and it was grand, with the dogs nobody broke in. They barked if anybody came about the place. The men were very good to my brother when he took ill. After my brother died in 1993, everything just changed after that, nobody took any more interest in the place until the council came on the scene and took it over. They hope to build a centre. I hope they do it up, I would like to see it done before I go.

Newforge Fine Foods
Matthew Neill and Dorothy McBride

Matthew. I worked in Mackie's making shells during the war, armour piercing shells they were. Anyhow, the Blitz discouraged me because the people who came to work, the girls who were displaced by the bombs in Belfast, they came to work in the factory, and the factory was a terrible place for them to work in. And they were really terribly distressed, it was a terrible job. I was a shell inspector but I gave it up when I noticed coming past one day, on my bike, Newforge Lane and the food factory.[85]

Newforge food factory.

So I went down and I went into Newforge factory and met a man called Fred Bevis. He was of Scots extraction. He was a very nice man. At first he didn't see any possibility of giving me a start at all because my qualifications were – I was a salesman, and really that wasn't what they wanted. They wanted a person to work in the factory as a machine setter. So he said to me, 'But you've no knowledge of machines, have you?' And I said, 'Well, six months ago I wouldn't have known the backside of a machine from the front of one, but I have been in Mackie's, so I know something about machines'. 'Well, come and see these,' he said.

Magazine advertisement, 1949.

And I went to see what they had at that time. Wee small canning machines, more or less hand operated, they were very small things. I said I could handle those and he said they were going to get bigger ones, we are going to have to send someone across to England and see the Metal Box Company who are going to supply us with the bigger machines. So, I got the job and went to the Metal Box Company during the war and I have vivid memories of that as well. When I came back I became in charge of all these new machines that came in. And gradually the job, of course, involved other machines like labelling machines, all sorts of machines.

So one day I was sitting in my, well, stokehole, really, covered in muck, and the factory manager came in and he said, 'You call yourself a salesman? Why aren't you out seeing Mr Hood about the job?' And I said, 'What job?' He said, 'Salesman!' So, when I was at my morning break a girl came along and tapped me on the shoulder, 'Mr Hood wants to see you in his office'. So I went up and I had to take some of the dirt off me before I went up into this palatial office. Mr Hood looked at me and, well, I came out with the job. And it was selling meat pies and sausage rolls.

The little place they had to make them was just a small corner of the factory and I was lucky because eventually, in about a year, the government opened 17 outlets in Northern Ireland for us! They were – what would you call them – places for refugees housed here – and there were 17 of them. One day in Liptons, I happened to be in, and they were preparing a list of goods to take to show these people that run these centres. And I said, 'Throw a few of those in and those in.' Sausage rolls and meat pies. And we got a contract for the whole 17. The result was that from being one person there was three of us, and from being a small corner of the factory it became an actual bakery section of the factory that employed about 100 people. It was 1943 to 46, something like that.

The factory was producing soups for the forces, canned foods for the forces, some of it meats, you know. A whole range of what you would call edible foods. But then after a while, after the war, they turned over to pet foods and it became a pet foods factory.

The old man Wilson wanted a way made from the factory out around the field and on out to Shaw's Bridge. So, the debris from air raid shelters, he said dump it here. It was called the Burma Road. That was his conception, he had a vivid idea of what should be done. I remember him showing me a field and saying one day we are going to have a car park there and the people in the factory will have their own cars here. At that time I thought this was nonsense because we didn't have any cars, just bicycles and feet! But yet the time came and that car park is still there. That was Clement Wilson. We christened it the Burma Road because at that time there was some talk about Burma; anyhow, we christened it the Burma Road.[86]

The people in Malone, they didn't want to know. The smells. Food being processed has sometimes not a bad smell but a smell. And the Malone Road people who lived close to it didn't want to know about it and especially before it became a factory for edible food, it had been the same kind of factory as there was at Burnhouse. For the first few years of the Wilson's factory it was a burnhouse. Dead animals. That would have been until 1939, maybe. But Wilson's was never greatly welcomed by the people of Malone.[87] To counteract that, Mr Wilson then opened the factory for people to come in to see that it was a nice place. He landscaped the grounds, that was his conception as well. I think he had an idea that a factory shouldn't just be a place that made stuff, it should be a place that people should come and relax in the grounds, not just for the staff but for people and it became a sort of a park. Mr Wilson, I remember him standing out on the hillside pointing out to – what do you call the man who looked after the grounds? Harry Langley. And he was standing on the hill and Mr Wilson was pointing out that and that – I want that and that. And Harry had to do that, get someone to do it for him. And he was a very good employer, no doubt about it.

They made frozen foods but it wasn't very satisfactory because they really hadn't the equipment and technique to be really good at frozen foods. They did make them and okay, they were saleable but they weren't up to the standards of some of the people like Birds Eye and so on. Not up to the standard of the very big people, they were more or less dabbling at it if you would like to put it that way.

Then they transferred the edible food business to Barrhead, Scotland and Newforge went to Spillers, it went to pet foods, they started to produce pet foods, cat food and dog food. The whole staff, the whole factory, was paid off because they started a new contract. Spillers took it over and eventually it was closed entirely. The Wilsons ceased to have any interest in it, the pet foods were taken over and I became a Spillers representative.

Newforge food factory. Sausage production, 1939.

Dorothy. Oh, most of my family worked in it some time or another, I worked for Newforge too, not in the factory, actually, but my sisters would have worked in the factory and my brothers worked in the gardens. My brother Hugh, he drove what they called the Scammell, a lorry with one wheel at the front – I don't think there is any such thing nowadays – he drove that and he would have been taking things round, you know, boxes of tinned food and stuff, to the barge on the Lagan. It was William Hewitt was the captain of the barge. He worked for Newforge. Newforge Fine Foods. Now, I can't remember what he done before the boat came along, but I think he must have been in the navy one time and this was why he was picked for this barge. He wore like a seaman's peaked cap. This cap showed that he was the boss! It docked behind the factory, up the Lagan and under the Red Bridge. It was the back of the factory. I don't remember it ever going further up the river, but I do remember it coming down under the Red Bridge and going to Belfast docks. There was only the one lighter that I ever remember, it was just something that was tried, it didn't last all that long. Tinned food, where it was being sent to, I don't know, but it was being taken to the docks. It had an engine in it, it was just like one of the ordinary barges, there was a cabin and there was an engine. Where Mr Wilson bought it, I've no idea. It was talked about and the next thing it was there.[88]

Mr Wilson, Clement Wilson – he gave his name to the park – he owned the factory. Father and him were very friendly and used to chat. He was Scottish. He used to come over to the lock house and chat. What they talked about – you weren't allowed to listen. Children were seen and not heard in those days. But he would have walked over, he walked with a nice cane, you know, one of those canes. A very handsome man. And he used to come and talk to my father regularly when he was here. Strathearn House in Newforge Lane was where he lived, but his main home was in Scotland. Newforge factory wasn't a burnhouse or anything then, it wasn't animal products. It was all soups and things. Strawberries and all were grown in the fields. As you go down Newforge Lane there is lovely house with a thatch called the Beacon; his daughter Helen Barbara Wilson for many years lived there. Across from that would have been the piggeries where they had pigs, and the other fields beyond that, right to Shaw's Bridge, were all strawberries and raspberries. When we were young, we used to go and pick them and get paid for it. And they had other fields somewhere else, you were taken there if you wanted to go. They had a cannery and they did soups, it was lovely, good. My family would have used the soup a lot. They had a product called Ulster Fry, it was smoked in a big slab, it was the loveliest stuff, you fried it. My mother would have bought a slab and it was cut like bacon. It was meat, processed. But it was lovely. The factory was operating before the war. Later Spillers took over and it became an animal food place. It passed on after that to another animal food place and it really went down after that. Nobody bothered with it any more. Now it is just a wee industrial estate, some of the old buildings you can still see, but most of it's all been knocked down.

Mr Clement Wilson.

Overleaf: Shaw's Bridge.

Shaw's Bridge and around the river to Drum Bridge

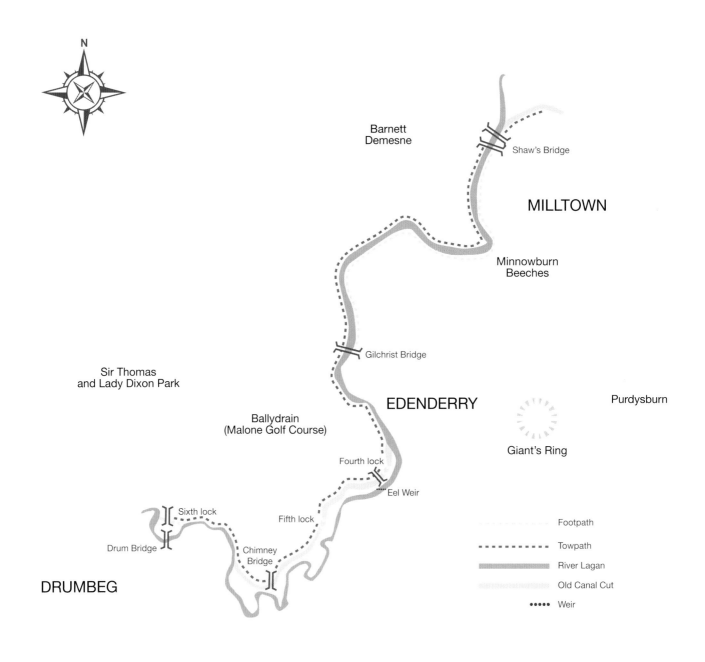

N

Barnett
Demesne

Shaw's Bridge

MILLTOWN

Minnowburn
Beeches

Gilchrist Bridge

Sir Thomas
and Lady Dixon Park

EDENDERRY

Purdysburn

Giant's Ring

Ballydrain
(Malone Golf Course)

Fourth lock

Eel Weir

Sixth lock

Fifth lock

Drum Bridge

Chimney
Bridge

DRUMBEG

Footpath

Towpath

River Lagan

Old Canal Cut

Weir

Shaw's Bridge and around the river to Drum Bridge
Lagan Lore

Passing upriver beyond Newforge and the third lock, the canal cut rejoins the river and the towpath leads to Shaw's Bridge. This is a long-established crossing point. Tree-ring dating of timbers recently recovered from the bed of the Lagan indicates that there was some kind of structure here in the early seventeenth century.[89] A map from around 1657 marks 'Shawes Bridge', and a brief comment published in 1744 that Shaw's Bridge 'was formerly small' suggests that this bridge was subsequently rebuilt or replaced.[90] Beside the five-arch Shaw's Bridge the river is crossed by a much newer concrete structure which takes the dual carriageway over the Lagan. This bridge, which has a single span of 58 metres (190 feet), opened to traffic in 1977.[91]

The landscape between Shaw's Bridge and the next road crossing upstream at Drum Bridge is dominated by the well-wooded grounds that were developed around large houses. The Legg (or Legge) family lived for generations at Malone, an estate at the end of the Malone Road on the hill above Shaw's Bridge. The present Malone House was built by William Wallace Legge around the 1820s. Subsequently, there were a number of different occupants of Malone House, and the last resident, Mr Barnett, bequeathed the house and grounds to Belfast Corporation. The property opened as a public park in 1946, and is now known as Barnett Demesne. Although the house was badly damaged by a fire in 1976, it has been beautifully restored and the grounds include spectacular wildflower meadows towards the Lagan and large old trees growing in extensive grasslands. This parkland landscape dates from the time of the Legg family, who is also still recalled in the local name Legg's Hill for the road from Shaw's Bridge to the Malone Road.[92]

Following the route of the Lagan upstream, several other large estates were created in the eighteenth century, including on the County Antrim bank, properties called Wilmont, Ballydrain and Drum. The original mansion houses at these estates were replaced in the nineteenth century, although all three of these Victorian houses have survived, and Wilmont (now a public park named after the last occupants, Sir Thomas and Lady Dixon) and Ballydrain (now a golf course) have retained their landscaped grounds largely intact. Sir Thomas and Lady Dixon Park has reached international fame through the success of rose gardens created here and, in addition to the annual rose festival, in recent years council staff have organised tours of a less well-known feature of the park, magnificent wildflower meadows that in the summer abound in orchids and yellow rattle.[93]

Over the years a number of other large houses were built in the area, including Longhurst, constructed at the top of a bank overlooking the Lagan in upper Malone at the end of the nineteenth century. This house, which remains a private residence, was the home of John Brown, son of John Shaw Brown, the founder of a weaving factory at Edenderry. John Brown was a partner in the family firm, but later in life retired to devote himself to scientific research. He was the first person in Ireland to own an automobile, a French Serpollet steam car, and he built a marvellous vehicle which he called an electric street boat or electric gondola. His concern about the state of local roads led to one of his inventions, the 'Viagraph', an instrument to measure road surfaces.[94]

Returning to Shaw's Bridge and crossing onto the County Down bank of the river, the road continues through Milltown, a small community that, prior to the building of the outer ring road, was spread out along a narrow country road. Milltown was hemmed in by the Lagan and by the private grounds of Belvoir Park to the north and Purdysburn to the south, extensive demesnes that were surrounded by high stone walls with entrances marked by large gates.

In the nineteenth century many people who lived in Milltown would have worked in Belvoir Park or farmed land leased from the estate, and the nearby Purdysburn Village was similarly closely associated with the Purdysburn Estate, which was owned by the Batt family. An article of 1860 noted that Purdysburn Village was entirely inhabited by people in Mr Batt's employment and provided a description of the cottages, which were said to be in the English style, each with a small garden behind and flower plots in front. Most were two storey, some built of stone and dashed while others were of brick. They had lattice windows, kitchens with tiled floors and a cottage range, a scullery with a sink and a closet under the stairs. There was also a school house 'under the superintendence of the National Board, the teachers being chiefly paid by Mr Batt'.[95]

Purdysburn was acquired by Belfast Corporation in 1894 for use as an asylum. The asylum utilised the former mansion house, and red brick accommodation blocks termed a 'villa colony' were constructed in the surrounding land. It was a largely self-sufficient community with workshops, churches and its own farm in which many of the patients worked as a form of therapy, but also to produce their own food. The mansion house was demolished in 1965, and little is known of the interior (it is said that some of the ceilings were adorned with the family crest of bat wings, which must have been unsettling for patients and visitors). However, the row of Mr Batt's attractive estate workers' houses by the demesne wall at Purdysburn Village, which were acquired by the Edenderry factory and rented to factory employees before eventually being sold as private homes, still remains.[96]

The construction of the canal above Shaw's Bridge in the 1750s created opportunities for the major landowners of the area. For example, the Maxwell family of Drum were compensated for making a track road and cut for the navigation in their lands at Drumbeg and in addition in 1756, Mr Maxwell advertised for sale land 'just by the Navigation of the River Lagan...very convenient for gentlemen Traders and others'. In 1775, Robert Stewart of the adjacent estate of Ballydrain advertised the carriage of goods of

Right: Shaw's Bridge with a staged scene of a man milking a lone cow in the middle of a field!

Canoeing in the Lagan near Minnowburn, 2009.

all kinds on his new lighter the 'Hillsborough Trader', and his neighbour William Stewart of Wilmont offered for sale in 1794 'entire or one half of her in partnership with the Proprietor', a lighter described as 'well calculated for the Lagan Navigation'.[97]

At Edenderry the full potential of the Lagan and canal to promote industry was realised. As early as 1780, there was a bleach green at Edenderry and by the 1830s, a water-powered flour mill had been built. In 1866, John Shaw Brown acquired the site and developed a weaving factory named the St Ellen Works. A mill village was constructed to house staff with three streets of two-storey terraced houses. The canal was used to supply coal, and water from a weir on the Lagan fed turbines to generate electricity for the factory. The proximity to the river did result in the factory occasionally suffering from floods, although the St Ellen Works flourished and its damask linen with its shamrock trademark woven into the cloth supplied many prestigious establishments. It is recalled as still working at full capacity in the post-war period, but by the late 1960s to early 1970s, it was employing fewer people and the factory shut in 1980. The houses were offered for sale to the occupants, and the factory buildings were for a while rented as industrial units, but have since been demolished and the land used to build apartments.[98]

Shaw's Bridge is today the busiest place on the towpath with visitors attracted by the extensive public parks, the development of canoeing facilities on the Lagan and the creation of walks. Downstream from Shaw's Bridge a short walk leads to a coffee shop and visitors' centre at the third lock; upstream, a stroll along the riverbank leads to the Gilchrist Bridge and the opportunity of crossing the river and returning on the opposite bank. As well as these short walks, longer journeys can be made on foot or by bike along the towpath, into Barnett Demesne or to explore the paths and minor roads around the Giant's Ring and Edenderry.

The Giant's Ring, which recent excavations have identified as an important Late Neolithic ritual landscape, has also been valued by people in more recent times. In the 1830s when workmen started to remove the circular bank, the landlord, Lord Dungannon, was informed and he directed that a circular stone wall be built around the earthworks to protect it. A plaque dated 1841 inserted into the wall commemorates his action and 'earnestly recommends it to the care of his successors'. In the summer of 1849, Lord Dungannon made one of his periodic visits to his estate lands around Ballylesson and used the large circular enclosure of the Giant's Ring to organise a tea party for his tenants.

> 'Within this enclosure, five handsome and capacious tents were erected for the accommodation of the guests; and in the centre, where there is a cairn, or pile of stones, a gaily ornamented flag-staff, with its floating banner, was planted. An immense gathering of the tenantry were assembled together, with a large number of ladies and gentlemen, who were invited to participate in the festivity.'

There were tea and cakes, dancing to brass and string bands, a harpist and also 'foot-races, leaping-matches, and all other kinds of rural exercise, enlivened by libations of cold punch'. The local newspaper judged the event a success and noted that Lord Dungannon 'mingled with the gay assemblage, evidently enjoying the happiness which he had created'.[99]

The Giant's Ring provides an ideal setting for events – it may have been created for this very purpose – and in the decades following Lord Dungannon's tea party the site hosted some other community gatherings. These included Christmas Day races in 1858 and 1886 and a Grand Bazaar and Summer Fete in 1923, which included in the programme some mystifying activities such as candle lighting and nail driving, and also washing and hat-trimming competitions.[100] Today the site is in state care and attracts visitors throughout the year, including at Easter when children gather on the sloping banks to continue the age-old tradition of egg rolling.

Returning to the Lagan and walking or cycling further upstream, the towpath meanders through an attractive landscape of farmland and old wooded demesne lands. Just upstream of Edenderry is the fourth lock, followed not far beyond by the fifth lock. Past the fifth lock, a small bridge (known for some long-forgotten reason as the Chimney Bridge) crosses the canal. This leads onto a strip of land between the canal and river where a narrow footbridge at the Eel Weir used to offer a way over to the far bank.

The next lock is at Drum Bridge, where the canal is crossed by the Upper Malone Road. This crossing, like Shaw's Bridge, is long established; William Petty's map of around 1657 shows a village at Drum Bridge and there is an old story that prior to the building of the stone bridge there had been a wooden footway.[101] In the late eighteenth and nineteenth centuries there was also a mill on the County Antrim bank of the river just above Drum Bridge, and the tailrace of this mill can still be seen, a long, still and dark pool of water that runs alongside the towpath and through its own tunnel under the Upper Malone Road.

Standing at Drum Bridge on 7 September 1763, you would have witnessed a magnificent scene. The canal had finally been finished as far as Lisburn, and Mr Greg, a Belfast merchant, took this opportunity celebrate the occasion with all the flair of a publicity-conscious entrepreneur. He took his lighter loaded with a cargo that included coal and timber along the canal and accompanied by his wife and invited guests, travelled in some style, making a special stop at Drum Bridge to welcome on board additional dignitaries:

> 'Mr. and Mrs. Greg had, upon the Occasion, invited a numerous company of ladies and gentlemen to make the voyage and to dine on board: the weather was fine, the prospect diversified with bleach greens breaking in at every reach of the river, together with the wood, lawn and meadow, and the happiness and jollity of the reapers almost in every field, cutting down the harvest, diffused joy and pleasure. They were met at Drum bridge by the principal gentlemen of the town of Lisburn, who came on board, and the whole company were entertained by Mr. Greg and his lady, in the most elegant and polite manner with a cold collation, and wines of all sorts in great perfection, besides a band of music, which played the whole way, to upwards of one thousand persons, who accompanied the lighter on the banks of the canal up to the town of Lisburn...'[102]

Mr Greg and his party would have been waved through the lock at Drum Bridge by the lock-keeper who lived in a two-storey stone house that still stands near the canal, a beautiful private residence in a wooded hollow on the far side of the road. However, a cottage that stood by the towpath close to the lock gates, and which in more recent times was used as the residence of the lock-keeper, has disappeared. The stonework of the lock is also now smothered by vegetation, and part of the canal channel has been infilled to provide a footpath under the Upper Malone Road.

The closure of the canal had become inevitable after the Second World War. In 1948, it was noted that there were just 20 barges in service and that only the transport of coal, together with some downriver traffic in food products, remained. The Lagan Navigation Company was dissolved, and responsibility for

River Lagan at Drum Bridge, 2011.

the waterway was transferred to the government, which proposed to shut the canal beyond Lisburn and see if income could cover costs in the section from Belfast to Lisburn. However, within a few years the entire canal was closed and most of the weirs on the Lagan were removed, allowing the river to return to its natural course.[103]

Although there are no longer lighters on the canal, and almost all of the old industrial buildings, tall brick chimneys and mill ponds have gone, some aspects of this heritage remain. Most importantly, the canal towpath, which for 200 years was used by horses slowly pulling cargos, is now a much-loved walkway and cycleway and is the vital link that connects many areas of public open space between Belfast and Lisburn. The entire region is now promoted as the Lagan Valley Regional Park, and large areas are now also in public ownership and managed for biodiversity and amenity. By looking at the landscape, finding out about local history and talking to people, the story of the Lagan Valley can be discovered.

I grew up in Milltown
Peter Johnston

Peter Johnston is my name and I was born in February 1938. I grew up in Milltown, about three houses down from directly opposite the Goat, Fairweather's shop, and went to school at Ballylesson and would have a little bit of knowledge about the general area.

The Goat pub had definitely gone by my time, I think maybe the Irish Temperance League bought out the licence for that establishment as they also did with regard to an establishment directly opposite Drumbo Presbyterian Church in Drumbo Village.[104] The Goat became a shop and I certainly can recall it during the war years, Mr and Mrs Fairweather owned it. Mr Fairweather was a professional golfer and his sister-in-law, Miss Ethel Patterson, served or worked in the shop. It was a general merchants in some respects and I can recall a petrol pump there that was in use directly after the Second World War. It was on what is now called the Old Milltown Road that goes to Ardnavally. They were also

The Goat public house. This building, on Old Milltown Road, is now a private residence.

agents for Ireland's or Johnston's funeral undertakers and they sold fruit and veg and that sort of thing. Fairweather's would have carried some vegetables, certainly, I can recall them having potatoes. Later they had a post office in the shop.

Owen's shop would have been between the Orange Hall Lane and Dunwoody's Lane, on the right-hand side going towards Purdysburn. I am not sure of Mr Owen's Christian name. They sold groceries and confectionery and they might even have had ice cream. On the other side of the road directly opposite Milltown Baptist Church, Alfie Downey and his wife had a shop, they would have sold cigarettes, sweets, newspapers and so on and would have delivered Sunday newspapers in the general Milltown/Edenderry area. Alfie Downey himself was a bread server, I think with McWatters' Bakery in Cromac Street in

Left: Edenderry Village and the factory buildings with a barge on the Lagan, c. 1920s.

Belfast.[105] Alfie was very much an entertainer and would have encouraged his daughters to sing, and his daughter, Patsy Downey, is well known in the music industry as being an entertainer, exceptionally good at playing the piano accordion, though I think it has got to the stage when maybe a piano accordion is too weighty for her to hold and she would use a keyboard.

William Black who had a horse-drawn cart, he would have sold fruit and veg in the area, a four-wheeled horse-drawn cart and he also had a barber's shop which in actual fact finished up being right beside Fairweather's shop on the Milltown Road. Prior to that he had a barber's shop at the top of the Milltown Hill, just round from where Lindsay's house was – Lindsay's had a big house on the corner of the Milltown Hill and Milltown Road. Milk would have been delivered by Greer's of Ballylesson, Tom Greer, Ballylesson. In a van. I can well recall him having an American-type van, during the war, and then he had a J-Type Morris, if I am not mistaken, after the war, possibly late '40s. One of his helpers was Bobby Caskey and another was Bobby Dunlop. You would have had a horse or pony and cart, two-wheeled cart, selling herrings and there were certainly, if I am not mistaken, two butchers, one called Brown from the Donegall Road and another one called Boyce from Sandy Row who would have delivered meat in the area. They would have come round on a Wednesday and a Saturday if memory serves me right. Calling with their customers.

Coal deliveries – there may have been a bellman or two about. There would have been coal delivered by the Antrim Iron Ore Company and a number of the other coal merchants from Belfast. A bellman is a man who would have come round maybe once a week with bags of coal, would have delivered a bag of coal to your home and you would have paid for it. The bags of coal in those days were ten stone – it wasn't a hundredweight bag – it was ten stone. The men had a leather jacket, and the platform of the lorry would have been high enough up to allow them to be able to take the bag and put it on their back and carry it in. There was a skill associated with carrying it. Carry it on your back then heave it right in, the bag over their head into the coalhouse or coal bunker or wherever. It would have been hard graft, now, hard graft. They would have been ten stone bags, more than a hundredweight. A hundredweight would be eight stone. I don't recall turf being burnt in the area. James Gray, who had a farm down the Milltown Hill, towards the stream, he would have sawn trees up for firewood but no, I don't recall turf to any extent being used.

The doctor would have been Doctor Nelson of Ballylesson. Dentist – there was a dentist called Monteith on the Lisburn Road. Doctor Nelson had a dispensary at Ballylesson which was opposite the shop at Ballylesson that was owned by Hutchinson. The shop is gone, there is maybe a postbox still in the wall of the shop, the doctor's surgery was directly opposite on the Lisburn side of Drumbo Parish Church, Ballylesson. Incidentally in Milltown, the right angles on the wall that, in fact, surrounded the Deramore Estate, is the house that Mrs Frances Weir lives in, and it was occupied by the district nurse; one of the

nurses was called Johnston, no relation of mine. The district nurse would have cycled around the area and presumably in association with the doctor would have visited patients that needed attention.[106]

There were two bus services, the number 13 bus service from Ormeau Avenue through to the Fever Hospital and on to Drumbo, and the number 22 service from Smithfield to Ballylesson and there would have been an offshoot of that to Edenderry. The 22 service would have been Malone Road, Shaw's Bridge, Milltown and what was referred to as the Low Road through to Ballylesson and to Ballyaghlis; some of the services would have gone through to Lisburn. They were run by NIRTB, the Northern Ireland Road Transport Board. Then, I think in 1948, it became the Ulster Transport Authority, at that stage they took over the railways as well as the bus services. The buses were a light shade of green and a lovely cream. They would have been 34 seaters, a driver and conductor and about 34 seats. I can certainly recall on occasions there would have been on the quarter to eleven service on a Saturday evening out of Ormeau Avenue as many as five single-decker buses to take the people home, which would have included people who worked at both the Purdysburn Mental Hospital and Purdysburn Fever Hospital. No double-decker buses were allowed over the old Shaw's Bridge – it was restricted to single-decker buses. Whenever the Newtownbreda Road opened up after the war – the Admiralty had the Belvoir Estate during the war and they had closed the road between Newtownbreda and Purdysburn – double-decker buses went up the Ormeau Road, through Newtownbreda Village and out to Purdysburn hospital. That would have been on a Sunday and Wednesday afternoon, for visitors to the Fever Hospital.

Harry Ferguson's demonstration driver was a man called Hylands, Hughie Hylands, and he eventually lived at the rear of the Masonic Hall in Milltown. Ferguson had a demonstration ground, it had HF on the gates, on the Upper Malone Road between Finaghy Road South and Dunmurry Lane. On the right hand side. The land would have been used as a demonstration field. I assume – and this is only an assumption on my part – farmers would have been brought there from the Balmoral Show in May of each year. The farmers all had Ferguson tractors, I don't recall any other tractors except Lowry at Purdysburn, they had a Fordson Major, the majority of the other tractors would have been wee grey Fergies. I don't recall seeing many horses ploughing. Certainly, I can recall seeing horses used, Thompson at Ballylesson had horse-drawn carts, rubber-tyred horse-drawn carts, drawing the manure away from his byres at Ballylesson and depositing the material at dung holes around the area, for spreading out over the fields at a later stage.

Thompson's stackyard, Ballylesson Road.

Thompson would have been the more forward-thinking and progressive farmer in the area. Gilliland and James Gray were big farmers, others would have been smallholders, like Willie McQuoid and Robert Burns. The main crops were oats, potatoes, barley, wheat. Hay – a fair bit of hay. I only recall flax on one year, on the land directly behind Fairweather's shop in Milltown, it was grown by a man called Duff, Old Daddy Duff, who had a couple of shops in the centre of Belfast and was a

Right: Threshing mill at the Giant's Ring, around 1955. On top of the mill are (L-R) Peter Johnston, Sam Copeland and Dee Murphy, owner of the mill. The mill was on hire to James G Gray, who had planted the land in oats.

butcher. I can also recall that field being used one particular summer that was very warm, for cattle herded out there to be held there until they could be shipped out by cross-channel steamer. And there was no water in that particular field, and those animals were kicking up quite a row. James Gray would have run 30 head of dairy cattle and he would always have had a Jersey or two to improve the butter content. At one stage he had pigs. The milk would have been uplifted by Emerson of the Castlereagh Road, Belfast. Thompson also had a dairy herd. I don't recall a great lot of sheep around the area and not a great deal of beef cattle about to my mind. Then there would have been the very large farm in Purdysburn Mental Hospital, my mother worked there before she and my father were married in 1936. James Gray only had a couple of labourers, and when a threshing mill came I would have been requested to help. Whenever threshing was taking place it would have been all hands on deck.

My mother had a poultry farm in that field directly opposite Fairweather's, and I can well recall as a child my mother had an incubator and would have brought out chicks. Whenever they would have been ready for moving on, day-old chicks, there would have been a notice put up on the gate, and the bus coming out of town would have stopped for some of the passengers to get off and come into our house and buy a dozen day-old chicks. The bus would have waited until they got the chicks and got back on. In a cardboard box, a shoe box, to rear for egg laying and to make a bowl of soup out of when they stopped producing eggs.

You know where Ballylesson Nurseries are? That property was owned by a man called Willie Moore, and Willie Moore had a big horse and a two-wheeled cart or trap which would have had rubber tyres on it, and when I say rubber tyres I mean wheels that would have been four foot six or five foot in diameter. And he delivered milk around the top of the Malone Road. Willie would have made his way home around midday, I can well remember in the mornings in the springtime my bedroom window would have been open and I would have heard the clippity clop, clippity clop, clippity clop of the horse coming down the Ballylesson Road. His son, in actual fact, became the engineer in Purdysburn Mental Hospital.

The first public housing I recall would have been in Gray's Park, that was on land that James Gray had sold to Hillsborough Rural District Council. James G Gray, who was the farmer down at the bottom of the Milltown Hill.[107] That would have been in the mid-50s. I can well recall the first refuse collection vehicle providing a service in Milltown, it would have been about maybe 1947/48. A lot of things happened directly after the war. We had electricity in our house when I was a child, and street lighting was put in about 1953/54. After that, Stewart and Partners started to erect houses in Belvoir Park, they owned Belvoir Park and won the contract to build the houses but then went bankrupt. There would have been 300 or 400 houses started maybe, but I don't know if they completed any of them. Later the estate was finished off. Another change around the place would have been the provision of water. Water mains and sewage: again a quare bit of that work was done by Stewart and Partners, before they went bust. Before that it was pumps and wells for water. The house we lived in was connected to a septic tank, and the water supply came from a hydraulic ram that pumped the water up to a tank which was at the end of the block of houses, directly opposite the nurse's house on the Old Milltown Road. This pumped water from the stream opposite the Masonic Hall. I never was able to work out how it operated. It pumped the water through a three-quarter inch pipe to the top of this tank, it didn't use electricity, it operated just by water pressure. It just went thump. Thump. Thump.

The new Shaw's Bridge was put in during the late 1970s. This new road came through the middle of Milltown, parallel to the Old Milltown Road. It cut through the Orange Hall Lane and Dunwoody's Lane. It went through Gray's land, McQuoid's land and Burn's land. Where we lived, a whack was taken off our garden, half the length was taken off our back garden. I remember my mother was compensated for the loss. Milltown had been an oasis of tranquillity in the '40s and '50s. The carriageway removed the rural aspect of the area and made it more urban. By that stage I was already married and had moved out to live in Belfast.

I can recall actually seeing the lighters moving up and down the Lagan and being drawn by horses. The horse would have walked along the towpath with a man with him. Come along to Shaw's Bridge and it walked up a ramp, across the bridge and down the other side. There are rope marks on the bridge to this day where the rope would have gone around or over. The man with the horse would have dropped the end of the rope down to the lighter whenever it would have gone on through the archway.

A couple of interesting points about Shaw's Bridge. During the Second World War, there was a gate on the centre of the bridge and if I am not mistaken, the cleat to hold the gate open is still on the bridge. There was a pillbox on both sides of the bridge. One where the ramp would have come up from the third lock, the second one as you would have gone down towards the fourth lock. There was also a series of concrete barrels – 40-gallon drums filled with concrete – that had eyes on them to chain them together, and the security forces rolled them right across the road. I think I am right in saying if you looked hard enough on the County Down side, the Edenderry side of Shaw's Bridge, you could find some of those barrels down there to this day.

I worked for a time in the old Works Section of the City Engineers and Surveyors Department in Belfast Corporation and would have been involved with the immediate repairs that were necessary whenever the side was blown out of Shaw's Bridge in 1972, I think it was, and I can chat about that at some length. There was allegedly 40 or 50 pounds of gelignite placed in a road gully and it blew the side off the bridge. I was involved with helping to get the bridge reopened for first thing on the Monday morning. It happened in the middle of a Saturday afternoon, beautiful weather in the midst of July. The Royal Engineers put a Bailey bridge over it and we provided temporary traffic lights at both ends. The bomb had been on the upstream side, towards the County Antrim end. You can see the different stonework where it was repaired. The day before, there was a bomb placed on a bus and it blew a hole in Queen Elizabeth Bridge, you could see right through to the water below. It was one of many explosions that took place on the Friday afternoon which became known as Bloody Friday.[108]

An interesting thing: the road from Malone Road down to Shaw's Bridge – the centre line of that road was the boundary between Belfast Corporation and Antrim County Council. There was an arrangement between the Belfast Corporation City Engineers and Surveyors Department and the County Surveyors Department in Antrim, that one authority maintained half the length of it and the other authority maintained the other half of it. A sensible arrangement. If I'm not mistaken, the concrete post which marked where one stopped and the other started may still be in place, on what is now the old road that runs next to the dual carriageway, in Clement Wilson Park. Then, whenever you got to Shaw's Bridge, the boundary between County Down and County Antrim was the centre of the bridge and there was a metal boundary post. That boundary post had been yanked out at some stage but you can still see the indentation in the stonework where it used to be, on the city side, halfway along, and there is a rope mark here in the coping stones of the bridge. The mark was made by the ropes pulling the barges. However, at the time when I started to work in the City Engineers and Surveyors Department, responsibility for the bridge was split three ways! Down County Council had one half and the other half was split between the city and county Borough of Belfast having one quarter and Antrim County Council having one quarter! This continued until local government reorganisation on 1 October 1973.

And I can give some local names of places around Milltown. Jacob's Ladder: the road from Ballyaghlis to Drumbo was so called because it twists and turns. The Three Mile Hill was at the top of the main Saintfield Road, where it levels out, at the junction with the Purdysburn Road. Presumably, it was three miles from the city centre. Back of the Wall: this was the name for Purdysburn Road where it used to be right next to the wall of the mental hospital going up the hill from Dixon's Corner. Dixon's Corner was the junction of Hospital Road, Purdysburn and Milltown Roads, where there are traffic lights now. You could have gone through the wall into the grounds of Purdysburn about 200 yards from Dixon's Corner, where the road branched off to Newtownbreda. Here there was a pedestrian gate with a pointed stone arch, it was called the Gothic Gate. The Low Road was from Shaw's Bridge to Ballylesson, the Cut was from Low Road to the Ballynahatty Road and the Sandy Pad was from Milltown Masonic Hall to the Ballynahatty Road. Purdysburn Hill was the road from Drumbo Parish Church to Purdysburn Village and Hospital Road, and we always knew the road from Shaw's Bridge to Malone as Legg's Hill.[109]

Everyone did something, Purdysburn was virtually self supporting
William Glover

A chum of mine, his father worked in Purdysburn and I think he was speaking to my mother and later suggested I might consider applying there, so I did apply and was accepted. It was temporary, for 18 months. That was in the middle of 1943. I was a male attendant. You were one of a junior team of between five to eight staff within any given ward with 40 to 50 patients. Your responsibilities were different from ward to ward, some were basic nursing just, some of them were getting the patients up at 7.30, they had their breakfast from 8.15 to 8.45, then they were allocated into several occupational squads as required. Some were employed on the farm, some employed about the workshop, some with the upholsterer and so on. Your basic responsibility was either to safeguard them when they were out or, alternatively, if you were on ward duty there was personal hygiene – baths a couple of times a week, shaving a couple of times a week, basic care. Staff came on duty at 7.00 sharp and you spent your first three quarters of an hour getting patients dressed and down to the dining area/day area.

William Glover, 2011.

There were two upholsterers in the hospital, they were responsible for doing the repairs to chairs and curtains and all that, bits of carpets here and there. Patient's clothes were bought in from suppliers but were repaired on the premises by the tailor. There were two tailors at one time and they did repairs to clothing. There were a number of patients, I would have thought at that time about six of them, in the tailor's shop. Some of them were people who were tailors before they had to be admitted to hospital. Terrific men, very capable of doing quite a bit, and relapse maybe for a period of two or three weeks and then pick up again. There was a shoemaker as well, and he had five or six patients helping him in the shoemaker's shop. You see, you have to realise that at that time there were nearly 2,200 beds occupied, roughly about 960 to 980 male and 1,100 female and there was also about 60 to 80 patients who were away at other hospitals during the war but came back in '45, early '46. There were tenders every year sent out for supplies for shoes and clothing and things, and in the workshops they did repairs.

There was a very limited number of women – I would have thought on the region of 15 or 17 – worked in a place called the laundry, they had their own laundry manager. A steam laundry. There were about six laundry assistants that were employed and some female patients helped them. Some of the women would be occupied in housecleaning and dusting in a limited way, two or three would be in the kitchen area, responsible for washing dishes, it was all hand washing.

If you were in charge of a working party, you saw that they had their coats and appropriate footwear and all that sort of thing. They had a tweed type of material suit and they had their overcoats if the weather was cold. They also had a working boot, a farming boot. They would usually be made up into different working parties, depending on what they would be doing. If they were tidying up roadsides, sweeping and that type of thing, the working party would have only consisted of about 10 or something like that. On the farm, there could have been anything up to four or five different parties going out from different wards.

There was a farm manager, a deputy farm manager, a man in charge of the pig yard and then I think at least four farm labourers employed. They grew all their own vegetables, their own potatoes, supplied their own pork. On a monthly basis pigs were taken to the abattoir. There was a small chicken farm. Horses used for ploughing and at a later date then, of course, they acquired tractors. It was self supporting in milk. There was a large Friesian dairy head and they kept two bulls down there. Down in the walled garden at Purdysburn House, it was a separate entity, it was a quarter to a half a mile down the road from the main hospital – you see there was about 500 acres all together. In the walled garden there were greenhouses and grapes grown. They produced tomatoes. There was a smattering of apple trees and pear trees but no orchard as such. They grew flowers and that sort of thing, cut flowers for the wards. A man called Eclan Young was in charge of the walled garden, and again, there were maybe eight to ten patients involved in it. In those early days when there was no central heating, all the wards were heated with coal fires and wood block fires. For example, the villas had four fireplaces downstairs, and upstairs in the wards there was two fireplaces. Now, after storms and that type of thing some of the patients who would have been working on the farm and were more able would have been formed into a group to go down and cut some of this timber.

The farm was still in existence, if memory serves me right, until the late 1950s. At that time there was a movement afoot for to do away with these farms. Personally, I felt at that time that the farms were a good thing because a large percentage of the people in the hospital came from a farming community. Those who wanted to do away with them would use very emotive terms, like slave labour and so on, but there was no slave labour. If a person could not do something they weren't forced to do it, or if they didn't want to do it they didn't have to. A lot of those people went out in the morning at 9 o'clock and were back to their villa at 12 midday and they had their midday meal and they went back out about 1.30 to 1.45 and then came back again at 4.30, so it was by no means a long day. Everyone did something, Purdysburn was virtually self supporting.

The patients went to bed at 7.00 in the evening but as years passed on and television came along, patients were allowed up until 9.00. Night staff came in to take over at 9.30 in the earlier days, that I remember. Later on they started at 10.00, and went off duty at about 7.30. In the early days your working week consisted of about 55 hours, you had three days on and one day off. Few of the staff lived in, most of the resident staff would be students or trainees.

You were given a bunch of keys – two or three keys on it – for locking both external and internal doors you had to open. There was a number on a little brass number plate and it was issued to you, a note was taken and you were responsible for those keys. The keys were about three inches long, substantial keys. In the early days you had to sign out at the gate lodge and you left your keys on a key board and collected them again when you returned. Each villa had 50 to 75 people, the female patients were in villas on one side of the hospital site, the men on the other side. Purdysburn House was also used for female patients, and the courtyard – there was female patients in that as well. Male staff had very rarely reason to be through the front door of Purdysburn House; as a matter of fact, when I first went there, there were demarcation lines. You would have been apprehended. If you were apprehended you would be severely punished, if not fined, because you were out of bounds on the female side of the hospital. I remember being stopped. That was in being a number of years after I went there.

There were two churches in the grounds at Purdysburn hospital. One of them was the Roman Catholic church which was taken care of by, in my time, the parish priest of Carryduff. The other, the Protestant church, was catered on a third Sunday basis by the Church of Ireland – the rector of Ballylesson, he took it, the Presbyterian minister – I think he came from Dunmurry, and a Methodist from the Ormeau Road. They took the services. The vast majority of the patients went out to church and they could have been visited in the wards by their clergy. All that was said was the charge nurse or whoever was in charge of the ward at the time would have said, 'Right, who's for church?' They are vandalised and used for

One of the two church buildings at Purdysburn, 2011. They were last used for services in the 1990s.

stores now. There was a working party of both male and female who went up every Friday to the churches for half an hour or three quarters of an hour to see they were dusted and done.

For recreational purposes dances were held weekly in the hall in the hospital, usually on Monday afternoon, there would be as many as 120 to160 men and women. In my early days the dances had a band of three or four staff. The chief male nurse, a man called James Gowdy, played the violin, it was his hobby, and a man called John Vogan played drums. Another played the piano accordion. In later days they used records.

In the early '50s, finances were made available to employ an entertainments officer and two female assistants. He was responsible for the organisation of all parties, trips outside to the seaside or whatever. You see, there were trips throughout the summer, a couple of buses would come and take away anywhere up to 40 patients up to Tyrella Beach or somewhere like that. And then as time wore on, groups of patients would be taken down to a show where there were complimentary tickets or even tickets were purchased, he got a budget and they could go to a show down town or anything like that. There was a sports day, it was an annual event. Around this time occupational therapists were employed and there was greater diversity of activities and it was seen as hopefully a step on the way to discharge.

There was very little medication available for schizophrenic-type patients in those days except for insulin therapy. There were one or two drugs for epilepsy, and there was a foul-smelling drug, paraldehyde, that was widely used to induce sleep for restless patients. Post-war they started to use electroconvulsive therapy for selected patients. On a limited scale, for the more disturbed, provided they were suitable, they used prefrontal leucotomy. From the mid-1950s you had a proliferation of drugs that improved medical care.

Later on we started to discharge them out into the community. The numbers in Purdysburn decreased. Those early villas, you see, when I went there first had up to 70 to 80 patients in them, spread over two floors, and when I left they were down to 36 or 38. A number of them are now closed, used for other purposes. Of course, it is right that they are out in the community, provided they get the backup and support.

I was promoted, the first promotion was in 1956. In 1964/65 I became the deputy chief nursing officer until further reorganisation took place. In the summer of '84 I resigned. I didn't have to resign, I was only 59, but I had the necessary service to give me the full pension. I was acting principal nursing officer at this time, you see, there were changes taking place every few years. To tell you the truth, it was becoming a wee bit onerous because you had all these new edicts coming out; you had to do this, you had to do that, you couldn't do that, you couldn't do the other thing. Looking back, it was quite enjoyable, oh yes, I think if I could do it all over again I would still do it, but you would see it possibly in a different light. I think it was run as well as it was possible to run it.[110]

A squad of patients went there every day
Essie Murray

I'm from Carrickatuke, in Armagh, I came to work in Purdysburn hospital in November 1953, in the dairy. Work wasn't just so plentiful around home, and a friend found an advert in the *Tele* and sent it up to us. Filled it in and came and had the interview and got it. I had been for a year to Loughry so I wasn't just leaving home. I was a dairy maid: keeping the dairy clean, the bottler clean, the cooler clean, bottling the milk and counting it all up at the end of the day and seeing that it was all there for to go out to the different villas. I hadn't a lot of bookwork to do, you just kept the utensils clean. It was machine milking. There were 75 milking cows in one byre and about 20 others, heifers and cows ready to calf. There were two bulls. They were all British Friesians and they were shown every year at the Balmoral and Saintfield Shows, it was a pedigree herd and they won prizes.

This was Purdysburn farm, in the grounds of the hospital, the byres were by the Purdysburn Road. The time of the year when the cows were down on milk they would have had to have bought in milk. Now Bamford's would have been the main one that supplied them, Bamford's Dairies. At that time we would have put out milk for about 1,400 people. That would have included doctors and nurses as well as patients, the milk went to the canteens, there were staff living in, a lot of staff lived in then. There were no sheep kept but they had pigs next door to the cows, there was a pig yard. They would have had horses, but the horses had all gone by the time I got there, it was tractors. They would have been working horses. I think the pigs were maybe sent away to be slaughtered, but they were brought back to the store as bacon. They used to have chickens, but before my time. The hospital grew its own potatoes, grew their own cabbages. Patients were used, a lot of the patients were used to harvest them. And then they had the garden where they grew their own tomatoes, lettuce, strawberries, I think they had raspberries, they had rhubarb. All the commonplace fruit and vegetables. Down at Purdysburn House, there was a walled garden where they had greenhouses and lots of the vegetables. A squad of patients went there every day.[111]

Essie Murray, 2009.

Sometimes the patients had a hard enough time of it, other times I think they enjoyed it. For a lot of them it was home from home for them. A lot of patients were there for, oh, for ages. Not in for a short time, in for years. There was one man he would have been shell shocked, he would have boxed himself when he seen his reflection in the window out of the dairy that was used as a hatch to lift the crates of milk out of the dairy to the fridge. And if the window was open, he would have started boxing, boxing the shadow, the reflection. He didn't do it all the time, he only did it sort of every now and again, whatever mood would have come over him. But he was very quiet.

They never were never allowed cigarettes, they were allowed tobacco. And I didn't smoke but always had to carry a lighter because you would have seen them coming, they would have got a wee tin box out, a bit of newspaper, put a wee taste of tobacco in, rolled it up and then they would have come and, 'Have you a light?' It didn't matter if you smoked or not, you carried a lighter. They could have set fire to somewhere if they had a lighter or matches. They weren't allowed either, you know. There was always someone came, and I suppose it was the same on the wards, the nurses would have given them a light.

All the villas would have been locked, but there was quite a few of the villas where in the afternoon nurses took out patients, maybe three or four nurses if it was quite a good day would have taken the patients for a walk round in the grounds. Female patients were on one side and the males were on the other side. When you went in through the front gates there was the admin offices and there was an avenue down to the left and there was one went down to the right. The men were in the right avenue and the women were on the left avenue. The villas are like big houses dotted through the grounds. It was nicely laid out. It was kept neat enough. It was kept fairly well.

There was one man and I don't know what his idea was, but he could have come down and there was a great big tank – a tank that caught rain water – it was in the bull yard, handy for to get the bulls a drink if they were out, and that ol' fella would have took that shirt off. And the next thing, if you didn't watch he had the shirt in the tank of water and he was washing himself. And then the shirt would have been back on him again. You couldn't have watched him! And he washed himself in freezing cold water even in the wintertime. Seen him breaking the ice on it for to do it. Whatever idea he had. Just some notion he had.

The violent ones were locked in, and I think there was a padded cell, in villa 11, wasn't it? Villa 11 never were out, but ones that worked on the farm or worked in the byres were never violent. Some would have walked around the roads, some of the people around about were afraid of them, but when you worked in the hospital you knew that they weren't going to do you any harm. There was one old fella and he used to go and he would have gathered fire lighters in the plantation, got a bag and put these fire lighters in and come and give them to you – wee twigs, bundles of twigs for lighting the fire. Everybody then had an open fire. Gathered them up and put them in a wee bag. He would give you this and you would maybe give him some pence and he went away. Then there was one fella, he came round to our door one day. My daughter was there and she went to the door and he said, 'I heard you were getting married, and there's a wee present for you'. And what was in the bag but a flat iron, for ironing her clothes! One of the old flat ones you heated on the top of the stove! She didn't use it, but she wouldn't let me throw it out, and nor when she went away she didn't take it with her, it's still here!

Purdysburn Village by John Maddock, 2007.

I left in 1956, after I had met my husband, he was a mechanic and was at the farmyard doing something, we got talking and I got asked out and that was that! I lived in the centre block of the Grahamholm, they had bedrooms there. Then we moved down into Purdysburn Village. It was a very nice, friendly place. It was small, yes, at one time there was only 21 houses in it – in fact, when I came number 1 was down, because we lived in the bottom house and it was number 2. My mother-in-law lived here then, it was number 21. When I came here most people in Purdysburn Village worked in Edenderry factory, the factory owned the houses. We were able to get a house because my husband Bertie's father worked in the factory. The old buildings had little diamond windows. We were led to believe that the houses in Purdysburn Village were originally built by the Batt family when it was a private estate. They lived in Purdysburn House before it became an asylum. There was a school, it is up on the road, not in the village, it's up, separate. If you are going up past Belvoir Park Hospital, the Fever Hospital, it's straight up in front of you, behind the hedge. That was the school.

I was never in the old Purdysburn House, I was outside it, but was never in it. Oh, there was patients in it, I can remember patients in it. Then it needed repair, and they built huts and took the patients out of Purdysburn House, and I suppose it was going to take too much to repair and then I suppose it would have been dangerous and so it was demolished.[112]

The Young Offenders Centre was built near Purdysburn House. Part of it is on a flat field where they used to have a sports day for the patients. They always had a sports day – egg and spoon races, some of them the tug of war, just all the normal things maybe you would see at a schools sports day. The patients would have been doing it and I think a lot of the staff would be watching. Some of the families, yes, I suppose did come, but some of them I don't know if they had any families or not. At that time, when the hospital started, if you had a mental disorder it was looked on as a disgrace, you know. There was

Purdysburn House. The mansion was demolished in 1965, though adjacent outhouses remain.

patients, ones that were around the farm, that never seemed to get any visitors, unless they had them at weekends or nights, but I don't think they did. I don't really think they did.

Then there was the old roller skating rink. It was down by the Back Avenue – we called it the Back Avenue but it could have been the main avenue into Batt's house – and it wasn't too far off the road and it was all concrete. When I came grass was beginning to grow in it. There were lovely wee huts at either end of it, like wee summer houses. It was down the road going in opposite the Fever Hospital, on the left-hand side, when you get down there into the hollow, on the flat, near the stream. I was told it was for roller skating, but I never have seen anyone roller skating on it. You could see it was neglected because there was lots of grass growing up round it. Oh no, no, it wasn't the hospital built it, it was Batt built it. There was also an icehouse in the grounds and there used to be a pond. Now I haven't been there for years and years and years and the DOE offices, I think, done away with the pond. It was the most beautiful pond in the springtime, they had lovely rhododendrons and the first time I seen it – I will never forget it. The water was still and there was a beautiful pink rhododendron – not the usual colour of the wild ones – and there was, I think, a white one and the reflection in the pond, it really was lovely.[113]

Edenderry factory owned all the houses
Derek Seaton

Well I was born and bred and lived in Edenderry for 70 years, but for health reasons the doctor said my house was not suitable and I had to get out of it, for the stairs were very narrow and awkward, you know. I have Parkinson's and I can fall very handy, it's the worst part of my trouble, I have the shakes fairly stopped. I have come here to Belvoir Park over a year now.

I was born in number 50, where my grandmother lived, my grandfather, the whole family lived there. There were all nicknames for the streets. Our street was called the Hen Backs because it was the only street in Edenderry that kept hens! That's all the name it ever got. Well then, as you come into the village beside the canteen or dining room as they called it, there were, I think, 24 houses and they call it the Front Row, facing onto the factory. Now there were two other rows of houses went up to the left when you got to the top of the Front Row and they were called the Half Crown Row because, for the simple reason, that that was the rent for them for the week, you see. Straight on was the Hen Backs.[114]

There were three or four different builders. There was a man called Ward, he built the second half of them, and I think it was a man called William Beers built the first number, I think he built the factory and I think the house we moved over to after that, 34. It and 35 were the two biggest houses in the village. See I was a foreman down there, then I got moved into the costing office. And that's how I got the house so quick, because them two houses were especially for foremen, you know. An ordinary weaver or winder, they weren't allowed that, we had that privilege.

Now they called that field at the end of the village that goes up the Giant's Ring the Grey Stone Field. There is another field with a name – as you are coming down the hill into Edenderry, where the shop is – they called that the Hall Door Field for the simple reason that Edenderry House was there. It was burned. It was massive – have you ever been up at Barnett's house? It was the size of Barnett's. And then the other field below that was called the Meadow Field and it was the only field down there that was worked, the rest of the fields, as I remember, were for grazing.

The houses were all built in different stages. Now the ones down at the Front Row, they were built in 1912. The oldest houses in the village were built in the 1860s. I know for a fact that my house was built in 1894. I have the dates and all, I made it my business, for I always had an interest in anything like that, you know. Some in the Hen Backs are not brickwork at all, they are all stone and plastered over, you wouldn't know the difference. Then above that, number 57 up 'til 44, you called them Lobby Houses. You had to go up this stairway and walk along to go into them. You had one family downstairs and another upstairs, but they didn't call them flats. Nobody can tell you much about them, this was a long, long time ago. The

Edenderry from the hill behind the village shows (L-R) a rear view of Front Row with well-kept back gardens, the recreation club, dining room, shop on the corner and some of the factory buildings, c. 1935.

house numbers in Edenderry were supposed to go from 1 to 105 but there are eight houses missing because some – they called them Cage Houses and I think there was eight of them – they pulled them down in 1929.

I will tell you another thing, if you went over there and you hadn't been in before and you were looking for a number, you would have a job finding it, I'm telling you, because they are not done in order. When you are coming in, the Front Row, as I call it, is 77 and when you come to the end and go to the next house it's number 15! You see, you would get confused! On the other side it's 57 and the top one would be, wait until you see, 36.

Edenderry factory owned all the houses. See if somebody was getting married, they might have had to wait six months to get a house, but say it was a fella and girl that were getting married and the both of them worked in John Shaw Brown's – no problem there, they got the house, you know. It was a whole different situation. I remember at one time our rent – four and nine pence our house was a week. Half Crown Row were the cheapest. They all varied, there was hardly any more than half a dozen in a row with the same rent. When someone retired they were allowed to stay on in the house. I told you about those

Cage Houses. If there was just the one and they retired and were fit enough, they were put into one of those Cage Houses because there was only one bedroom in them. They were by the Front Row, down at the Lagan, just up at the very end of it. There is no one that can tell how they got their name, but as far as I can gather they weren't just ordinary roofs, they had a half-moon shape. I think that probably they were just nicknamed that, you know. As I told you there, they had special names for everything, for the different houses. They were pulled down and I can tell you the year, 1929. I don't think there was any more than two allowed to live in them and there was nobody young got them.[115]

The houses had no electricity until 1935, and I have a good way of proving that to you, for it was the year I was born and my mother always told me that they had to use the oil lamps, and apparently for the first six months I could have kicked up a bit of a racket, you know, crying and stuff, and they were fed up lighting oil lamps. And she said that after about six months they moved in and put the electricity on and that after that there was never a mute out of me!

My grandparents both worked in the factory. If you go down the Lagan, there was a place there they called the Honeycomb. I don't know how many houses there was but our family lived there before they come to Edenderry. There was a special boat there that belonged to John Shaw Brown's son, Jack Brown. They had a boat there. See Jack Brown he owned what they called the Honeycomb houses and there was a boat and anybody from the Honeycomb who worked in the factory got the boat there and back, just across the Lagan. The boat worked with ropes, whatever way they done it. It was gone before my time. They were my father's people that lived at the Honeycomb, John and Mary Ann Seaton, I think they had 10 children, but a couple of them died you see, when they were young. I don't really know how many other people lived there, but have you ever heard tell of Elwoods the funeral undertakers? I knew them extra well and Jimmy Elwood's grandmother lived next door to my grandparents and on the other side lived the Copelands. They all worked at the factory. I don't know why it was called the Honeycomb. I don't remember the houses standing.[116]

I left school at 14 and started working at the factory and also went at night to the Belfast Tech for four years, three nights a week, to learn linen weaving. I worked most of my life in the winding department. My father had other things lined up for me but it ended up that I got a good position out of it, you know, and I didn't want to leave. You were either a weaver or a winder. The foreman, they called him the Winding Master – funny names in them days – and I was his assistant, I had to give yarn and stuff to the winders, oil the machines and stuff. When they saw how good I was at figure work, it wasn't too long until they had me up into the costing office.

John Shaw Brown's had five golden medals and there wasn't another factory in Great Britain ever attained five gold medals, you know they done the finest damask in the world. It was a quarter past eight in the morning 'til six at night, it was a 45-hour week, but when the place closed down, I think it was 39 hours. There was a factory bell. After the factory shut, someone removed it. They must have had a crane or something at it, for it was a fairly heavy bell, you could have heard it, they tell me, over at Drumbo and Purdysburn. That bell was ringing at twenty minutes 'til eight to warn you it was time to get out of bed, get your clothes on, and then it went off again, that was the last call to giving you five minutes to get to work and clock in. If you were late, it was deducted. You had three quarters of an hour for lunch.

There was a canteen across the road from the factory where you could get what they called a mug of tea. I never used it, I always went home. The shop in Edenderry was next to the canteen. The factory owned it. Mrs Field ran the shop and she run it thoroughly. There was no messing about, if you went in and wanted a pound of sugar or packet of cigarettes, they were handed straight over – these country places half the time you went in they never had what you wanted. She always had everything. There was never a pub, they would usually go over to Bob Stewart's or the Homestead. The Homestead is pulled down now, a real shame.

It all dwindled down and at the end I was actually doing my own job in the costing office and running the winding shop at the same time. In 1980 the factory closed. They just came down and handed me an envelope. I can still see the old folk with tears, they had worked in it all their life. I got redundancy out of it, but you see all the old women, they didn't get tuppence because of the simple reason that they were over 65. I felt really sorry for them. Shut just all of a sudden.

People at Purdysburn Village worked at the factory and I will go a step further than that. You are familiar with Carryduff? There was old people, could be coming up to 70, and they worked in it six days a week and they had to walk it there and back to Carryduff. There is a fancy name for a particular area up there, they called it the Duck Walk – a whole lot of people up there could tell you that, you know, the Duck Walk. Like, could you picture yourself what time you would leave the house to walk from Carryduff and they had to start at seven o'clock in the morning.[117]

At Minnowburn Bridge there was a gate and two pillars and I think it was a bus or a lorry hit them, but the gates had already been taken away, about 1920 or thereabouts. The gate was locked at nine o'clock every night. Then when you came up about the farmyard there were more big gates. The farmyard at Edenderry House. The boys used to go out, maybe to have a drink. But they knew, they had their heads screwed on, they could get back no problems. But it just shows you, like what sort of people you were dealing with!

The road from the Minnowburn up to Edenderry, it was called the Avenue. Somebody went and stuck a notice up now, Edenderry Road. That's a lot of nonsense, it was the Avenue. When you came to – see we

Derek Seaton on his electric bike, recalled by his wife Sally as 'his pride and joy'.

had special names – what we called the Hearty Trees, a big group of beech trees, there was a gate there and there was no one allowed into that, that was the entrance up into the grounds of the big house.[118]

Up the river, about halfway to Drum Bridge, there was a building near the towpath with a waterwheel on a stream, I remember it going round. In my time the Dorans lived there and it was a laundry for Ballydrain Estate. It's gone now. You can still see a little bridge on the way to Drum Bridge, a bridge over the canal. We always called it the Chimney Bridge, I don't know why. A man I knew used the bridge to take cattle over to the land between the canal and river, the middle banks.

We all went to the factory, every one of us

Sisters Susie White and Annie McCarter, their niece Marion McWilliams
and her husband Alan McWilliams

Susie. My father worked in JP Corry's, the timber people. That was where he worked and he went on a bicycle every morning, and when the bike came home at night it was brought into the kitchen, and the carbide light was filled and the tyres all examined so everything was ready for morning. My mother was a weaver. We came here to Edenderry in 1939, it was the day the war started. She had to go and see the manager and ask him about a house and he said yes. It was that first house – as you go into the village, that first house there, that was the house we were coming to – but then the manager come to my mother and said, 'Mrs Mackey, would you mind if I give you a different house?' The man in the office – Tommy Taylor they called him – he was getting married and he wanted the first house, and he had a good position in the factory. My mother said, 'I don't mind what house I get as long as I get a house.' And then that's why we went up the Big Street. We were number 30. The Big Street is the street I live in now, it's straight on right to the top of the village. We lived on the left. And then later on we moved over to the bigger house on the right. And I'm in the house now that my mother was in, number 43.

Front (L-R) Susie White, Annie McCarter. Back Marion McWilliams, Alan McWilliams, 2011.

There was nine of us, seven of a family and father and mother. Three bedrooms. And then when we moved to the bigger house there was an attic. All the girls were in the attic and the two boys were in the back room, my mother and father in the front room. To get the house you had to work here. All the houses were owned by the company. The manager he said to my mother, 'Would you promise me some of your family?' She said, 'I will.' And she done that. We all went to the factory, every one of us. Jane was the first to start, then Martha, John, Annie, myself, Margaret, and Bill was the last to go in. We all went in. Bill was in the cloth room at that time, before he left for to be an apprentice bricklayer, and John was in the mechanic's shop, he was a mechanic, he looked after the looms; anything broke or anything, they took the part to the mechanic's shop. Jane was a weaver – she worked in the damask end. Martha worked in the damask end. Annie worked in the office and I worked in the weft office, I was in charge of giving the weavers their weft. And I had to give the winders what work I thought I would need for the weavers.

Annie. We went to school at Ballylesson. There were a fair amount of children. We walked up the Ballynahatty Road. I suppose about twenty minutes or so. There was just two classrooms, the big one and the small nursery one. That school teacher for the primary ones lived in Edenderry. Mrs Taylor. I left school at 14. No, I didn't go straight into the factory. And I had this cousin who worked in the ledger office and she was leaving to get married. And she came up one day and she said, 'Annie, would you like my job?' I said, 'Your job? I couldn't do your job.' She said, 'You could, you could.' That was 1945, and she came up two or three days after and she said, 'Annie, Mr Orr wants to know could you come down for an interview'. Interview! I was a bit on edge, you know.

So, my cousin stayed for a while, and then I got the routine of the job. I loved the job. I was doing the weaver's wages. The cloth was measured in the cloth room. The office I had, I could see them measuring the cloth. And then they weighed the cloth. I got the tickets with the yards and weight on them. My job was just the weavers. I put all that in the ledger, you may guess how heavy it was, the boys had to carry it over to the pay office! They copied that for the pay sheets.[119]

I lived in Edenderry until I was getting married in 1964. Then I moved to Grand Parade. I said to Mr Orr, 'I will have to give in my notice because there is no bus here that suits me coming out in the morning.' He said, 'No problem.' He had moved from Harberton Park to the Old Holywood Road and he said, 'I can come round past Grand Parade and lift you'. I worked in the factory until 1965. In 1992, I bought this house in Edenderry – my husband had died, it wasn't the same and the two boys were out to work. Oh yes, I like living here now. Happy memories, yes.

Susie. I enjoyed growing up here, we were happy. We had a good father and a good mother. Our mother made most of our clothes, baked all our bread, things like that. Real homely. We just played on the streets, played skipping, got tennis rackets, played tennis. We played marbles on the street, on the wee crib on the street.

Our mother was working in the factory when we were small. Our father worked in Corry's, he went there on his bicycle every morning. He left early, early. He was away before we got out of bed. When I started working in the factory my mother had left. It started at eight o'clock, eight to six. Five days a week. And then we got our lunch, we come home for our lunch. The bell rang early to waken the people and then it rang again at eight. There was a clock and you had a number and you had to punch that clock every morning going in, punch it coming out at night. If you were late it was taken off your wages.

There was a shop in the village but we had a man that came, Mr White you called him, he had a wee shop at the Ormeau Road, I just forget the name of the street, and he come around. He had lived beside us when we lived at the Milltown, and then when we came to Edenderry he came a day in the week and took my mother's order and then brought it the next day. That's where we got our groceries mostly. We were the only customer that he had in Edenderry that I know of. He had a

Susie White in the winding shop at Edenderry factory standing at the winding machines, a bobbin box beside her, c.1946.

van. He came round and got the order. The butcher came to the door once a week, and a fish man came too, mostly a fish man came on a Friday. You would have heard him shouting. And then there was a vegetable man came. Dennis Graham, he came round on a lorry with vegetables. When the shop in Edenderry became vacant he took it over. He was in it for years, Dennis sold everything. Going back Mrs Field had the shop, that's going back a long time. She had to make the dinners for the bosses. They went over to her dining room, they didn't come into the dining hall with the workers, they went into her dining room. The company owned the shop and that was the condition for her to get the shop, that she cooked for them. In the dining room of the house. They went round the corner and into the hall door.

Oh, there was indeed a lot of children in the village. Well, where the play park is now, there was a big shed – just a roof and they had swings inside it. We called it the tin shed.

Annie. You know where the new estate is up there? Were you ever up there? Past the Square, at the back of the Big Street. Well, we played up there. It was gardens and allotments for the houses. Remember we made wee houses of our own.

Susie. We made our wee house, like. If anyone had bits of linoleum that anybody was throwing out, we put it down, on the grass like, and got wee bits of delft. Set them up like a wee house. The nicest wee house!

Annie. Oh yes, I remember boats on the canal. Pulled by horses. They delivered coal to the factory. Bertie Waterworth was there. He lived in the village and he worked there, he was there every day and then when that boat came up, he was there to get the coal off. There was a lock further up. Tommy Singleton lived there. It was straight across from the locks, sitting up the hill. Where the golf course is now. A whitewashed house. A right size of a house. I was in it. Because the people that bought it afterwards had a party in it one night and we went up to the party. Milligan. Do you remember, the girl Milligan?

Susie. Tommy Singleton only had one arm. He was a wee small man. I didn't know much about him, he never was in the village. Just to come to get the bus and get off the bus. He had no family. He just lived on his own.

Annie. It must have suited him all right, he stayed there anyway. He was there a good long while. Didn't he fall into the Lagan?

Susie. He fell into the Lagan one night. And he drowned. They said that he had a goat up there and he went to take up the stake – it was in the ground – and he tumbled over. They said that's what had happened to him. Because seemingly, when the police came the goat was running loose. He had got the stake out but just tumbled in.

After that, the man Tom Rea came. He was a big man, a big tall man. I think he was on his own too, I don't think he had any family either. I remember seeing boats, horse drawn. There was a woman, it looked like a family on one boat. I mind a wee woman and she had children on that boat. They must have lived on that boat, I think. I remember her well. The boats were just black. No, we were never on them. But we used to go over when we seen them coming, we used to go over and watch them. I remember them leaving off coal, yes. I suppose about every couple of months they would have come to the factory with coal.

Annie. Oh, every three weeks. It took some coal to keep those boilers going.

Susie. The baths? I don't remember them working. They were down there just facing the Front Row at Bunting's Bridge where you go down to the Lagan. There was two of them, two set together. Number 1 and number 2, that's why the houses start at number 3. They were like wee sheds. I remember that. Now Sadie, she remembers you had to book to get a bath. You know, you had. In our house we had a big long tin bath and it hung out in the backyard. And then it was brought in and we had a boiler, a Burco Boiler. And that boiler was filled when anyone wanted a bath! Then as time went on we had a sister-in-law lived in Purdysburn, she had a bathroom and we thought this was great, we went over there! Then Annie went to Grand Parade and she had a bathroom and we went there! We never had a bathroom, so we never missed it. Different if we had a bathroom and then come from it, you know. But times have changed.

The Browns owned the factory. Mrs Brown, I remember her, I don't know what her first name was. She was a tall woman. How many Browns was there?

Annie. I think there was three boys.

Susie. I know there was one in Canada. There was a Richard Brown. I didn't know them, the only one I knew was George, like. No, I never seen them round the village.

Marion. George Brown was never about the factory; I mean he never was there until the warehouse on Dublin Road went on fire and they come out here and then George Brown came with them. Before that we would never have known who George Brown was. He had been working down there. It was Mr Shannon, Mr McCammon and Mr Orr. Mr McCammon, he was the head man. Then there was Mr Orr, Sydney Orr, he was in charge of the office, he took to do with all the office work. And then there was Frank Shannon and he was the manager, he looked after the factory. That was the three of them.

Annie. But we used to go down, May Day, to collect for a wee party. And she opened the door and gave us something, Mrs Brown. We collected for a wee party, that was all.

The fourth lock with Edenderry Village in the background, 1927.

Susie. There was always the May Queen.

Annie. I think the neighbours give us something to dress the May Queen.

Susie. Something in her hair and all. Wee flowers.

Annie. And Mrs Brown was an awful nice woman like, coming to the door and all.

Susie. She was a gentlewoman.

Annie. And you remember the night the war was over? Remember the Square. Dancing in the Square. Bobby Johnston got the piano accordion out. He lived up the village. The dance, nearly everybody was at it.

Susie. Up to the Giant's Ring? Oh yes, every Easter, we never missed. Took our hardboiled eggs. Painted faces and all on them. You couldn't have got sitting down for the crowds that was there! They came from everywhere. And they brought all the children, you know. Oh, we went up quite often to the Giant's Ring.

Annie. Well, in the factory I was in the ledger office. There was a pay office, that was where all the tickets were made out for the pay and then there was a costing office next to that. And there was the typist's office, and outside was the general office, three girls worked there. Mr Orr I was telling you about, his office was beyond that. And then the top man, Mr McCammon. Well, there was a boiler house, then the mechanic's shop, the carpenter's shop. What else? The weaving shop. Winding shop. Dressing shop. Yarn store. Cloth room, where they inspected the cloth. In the weaving there was what we called the damask end and the linen end. All the big cloth was in the damask end and tea towels and things like that in the linen end. Damask is when you have a pattern in the cloth. Tablecloths and did they do sheets? I think they did sheets, too. And they all had the shamrock on them, on all the linen. Sheets and pillowcases and napkins.

Susie. There was the card room, where they cut the cards for to make the patterns for the cloth. And there was a girl worked in the card room. She had an office on her own, but it was in the card room. Miss Gray you called her. And she done the painting, she planned all that, the patterns.

They started to call meetings, you know. And then – we just thought there was something happening. They said the work was very scarce. The bosses came and talked to us and then the next thing it was closing. They had a place in Canada, a warehouse in Canada, the bosses went to Canada from time to time. But seemingly, I suppose through the years, man-made fibre and all came in. People didn't want the linen, it was too expensive. Everybody was downhearted. Could do nothing about it.[120]

It was about the late 1970s when the Housing Executive started to do up the houses. It was Rodgers the builder, seemingly, he took over and done all the work for them. They took our back bedroom to make a bathroom. And some of the attics – where they had two of a family, two children, they divided the attic and made two wee bedrooms. But I didn't need mine done so I have the big attic. They put the back boiler in so you would have hot water. After that they asked us to buy them. If they would have sold them to us before they had done anything we could have made a better job of it. In my house we could have built a working kitchen and put a bathroom on top, above it, and kept the bedroom, but they wouldn't let us do that. They done the renovations and then they decided we could buy it if we wanted it.

Alan. My mother was born in Edenderry Village. She worked in the factory. My father was born at Purdysburn Hill, that's in the Ballylesson area. They finished up in Edenderry Village, they spent most of their life here. I grew up in number 96. There was five of us, five boys. I was the last one. I don't think Clifford worked in the factory, the rest all did: Victor, Bobby, Bill, myself. Bobby worked in the office. Bill worked in the dressing shop, Victor worked in the card room and I was at the tenting. It was good times. Looking back on the mill, to be honest with you, it really was a bit of a disaster. But at the same time everyone was happy and friendly.

Susie. The best of friends.

Alan. You could have got jobs in Belfast, but the time you were travelling, the bus fares, you were really no better off, you know. There were only two people I remember losing their houses. Both of them worked in the mill and they were the only ones in the family that worked. They left the mill, both of them left, so they had to leave their houses.

Marion. As long as one member of the family worked in the factory they wouldn't have put you out of your house.

Alan. I was born in the village, then school at Ballylesson, into the factory at 14 years of age. It was rough, the mill was rough. The working conditions. Especially for a tenter. Up in that roof, the glass, you know, the sun coming through, it would have melted you! You were looking after your looms. Whenever they had problems with the weaving, whatever alterations you had to make for them to get the cloth right, that was your job. Making sure the cloth was coming off right and the looms were going right and well oiled, greased. In those days you served five years' apprenticeship and one year improver, improver's year. Six years in total. The machinery they had, even in those days, was outdated, wasn't it like. Outdated. It was all old-fashioned.

There was six shares in the main factory. And 60 looms per share. That's what they called it. A share of looms. A tenter looked after his share. So he had 60 looms to look after and so many weavers, on average four or five looms each. Oh, it was noisy. Dusty. Very noisy and dusty. There is no way that would be allowed to happen today. Health and safety. It was dangerous. You had to be careful with them. If you were up, sometimes up in the machine, up above, you had to be very careful there, too. Bobby Roy lost a finger. You were apprentice to two tenters and then you had to attend Belfast Tech.

They were just a disaster to tell you the truth. Then they got a lot of new ones in from Monaghan or somewhere. A mill closed in Monaghan and they took them over but sure they were worn out too. This new manager came in and he had this great idea to buy the looms. They were supposed to be ultra modern but they weren't. They were played out, worn out too.

Susie. They said that was the start of the downfall, bringing those looms out.

Alan. Most of my life I was in number 96 Edenderry. That's up the Wee Row next to the gospel hall. Normally it's called the Wee Row but in those days it would have been called the Bride's Row. That's the first house you got when you married! They were the smallest houses and whenever you got married, that was the house the factory would have rented you. I only knew the front street as the Front Row and the Big Street, that's the one that goes straight on, then the Wee Street. Three streets.

Things started to go bad, they were talking of three days' work a week. I was actually made redundant at one time and then they changed their mind and kept me on. And shortly after that things started to go really bad in the mill. There was no future, it was obvious there was no future, and I got a job in Spillers Foods in Newforge.

All my life in Edenderry. Oh, it was great. You couldn't have beat it. You could still go to the cinema from here. Get the last bus home – that was a famous bus, the last bus to Edenderry on a Saturday night!

Marion. In them days you didn't need friends outside the village because there was plenty of people here.

Alan. See, there was the recreation club here, just here. And you had billiards, air gun shooting and all sorts of things. There was a football team here as well. St Ellen's football team, they played in the amateur league, and Ballycairn Presbyterian team in the church league. Quite a lot of the people in the village played in both teams.

Annie. Wee dances on a Saturday night.

Alan. Oh, it was a great place to live then. You could have went on your holidays and left the front door open, nearly. Definitely a good place to live then. It still is a good place to live. But it's changed now. There has been a lot of development, new houses where the factory was, the factory has gone.[121]

They called the village the Cage at that time, because there is only one way in and one way out. Up at the Square there were houses built like a cage, apparently that was originally where they got the name from. They were gone long before my time. There were gates at the Minnowburn and at the top of Hazley's Hill. I don't remember them shutting them, I don't mind that, but I remember the gates being there.

Susie. But they did shut them at night, they did.

Alan. I remember the barges. They delivered the coal, yes. I remember Tommy Singleton drowning. I mind that and then what do you call the guy who come after that – Tom Rea. I mind him, he came after that.

Whenever the barges stopped, the lock-keeper's house was taken over by the newsboys' club from the Falls Road, and they used the house for weekends. Malone Golf Club, they developed the whole course there, so that was demolished, taken in to be part of the course. Lindsay and Rosie Morrison lived at the next lock house towards Drum Bridge, but he wasn't the lock-keeper, at that time Tommy Rea looked after both locks.

To get over there you would have gone down over Bunting's Bridge, as we called it, and right on up the path and over the Eel Weir and then straight down over the lock. There was a river flowed underneath Bunting's Bridge, we called it the race, it came down here to drive the turbines. The water came off the Lagan at the high sluices away up Ballyaghlis, and the dam, that was a man-made dam, right down this side of the Lagan, it came down here to drive the turbines. They were going right to the end. Electricity for the factory.

I think they knew it was coming. At the end there was a lot of elderly people, very few young.

Marion. When the warehouse on the Dublin Road went on fire they all come out here. So there was a lot of strangers and all at that time. But they were mostly elderly people that worked within the factory.

Alan. Everybody knew each other and they all rallied round each other, helped each other. But sure, there are very few left. You don't know three quarters of the village now.

What happened to the bell? That's a good question.

Susie. We would like to know that.

Annie. We don't know who took the bell.

Alan. That was the bell that took the workers in. Whoever was the boiler man, that was his job. Quarter to eight in the morning he rang that bell. First bell at a quarter past seven then a quarter to eight, was that it?

I was born on 30 March 1905, in a thunderstorm
Auntie Girlie

I was born on the Malone Road in a lodge, Longhurst lodge. My father was the first man to drive a car for John Brown of Longhurst who was an inventor, he built Longhurst. My father was with him until he died. He was a chauffeur to John Brown. He served his time as a mechanic in Dunmurry mill first. Well, he came out of Dunmurry mill as a manager. He was to go to Comber and his mother kicked up a row and wouldn't let him go – she was an awful woman, she was very determined – she said if he went she would go to her bed and never rise again. So, he should have called her bluff, you know, but anyhow, he went then to John Brown, and John Brown was an inventor and John Brown had the first car and my father drove the first car. And it went with coke, it burnt coke.[122]

Auntie Girlie (Susan Irvine), 2009.

I was born in the lodge in Longhurst, a gate lodge, it's still standing, it's still there. I had two brothers and a sister: Percy and John and my sister Eileen. I was born on 30 March 1905, in a thunderstorm. The doctor later told me it was the worst night in living memory, a terrible storm, and he had lost his way! The doctors rode bicycles then. I went to Drumbeg school, I had two mile of a walk, I walked there hail, rain, snow and blow. And in the wintertime when you went in, there was always a lovely big fire. It was quite a big school, the Charlie Memorial. Longhurst was a big house, it was a lovely big house. He had two children, a son and a daughter, Eric and Rosalind, I played with them. The house is still there.

My father and I walked the Lagan, I was the only one that walked with him, the others were too lazy or always had something else on. After dinner on Sunday he would say, 'Let's go for a walk', and we took the dogs, we had dogs. We had Rags, Nellie and Gypsy. Gypsy was a retriever, Rags was a cocker spaniel. I remember the canal, I went to Drumbeg school, you see. And over Drumbeg bridge, the Lagan was there. And we used to stand and watch the lighters. And we used to go down the Lagan side, down the towpath. And we went and got on the lighter coming up and then we went up and then we were able to get off before it went through the bridge. They had a horse, sometimes there was a horse that pulled the boats up. And sometimes they were motorboats and they didn't need a horse. Some of them were quite good to their horses and some of them beat their horses. I used to be quite indignant, I remember being quite indignant, I think we nearly beat the man! This horse wasn't able to pull the lighter, the horse fell two or three times. And we lit on him. There was five or six of us. I think we could have thrown him into the Lagan!

Mr John (Jack) Brown, FRS at the wheel of his Serpollet car, c. 1896.

Oh yes, the lock people. I only knew – what was that you called him – a wee man and his wife – Theo McCord you called him. He was a wee man with a wee white moustache. That was Drum Bridge. They had a shop, they had a sweet shop, yes. You could have bought a pennyworth of Black Lumps or jellies called Jube Jubes or anything. Black Lumps? Oh they were just black lumps! Black Lumps and Bull's Eyes, a hard sweet and they were black, but they lasted quite long. And then there were Bull's Eyes but they were striped. Black Lumps and Bull's Eyes. And toffee. I never smoked but my brother John, he had bought cigarettes one day and he was standing in the kitchen at home and he had a fag in his mouth, and my daddy came in and he gave him a whale. John said afterwards to my sister, 'You know Eileen, I hadn't even it lit!' But he took the strap to him for smoking.

Then there is a house across the road, it's in Reade's Estate and it was used, you see, as the Lagan cottage but then the Miss Reades – it was on their ground – they built the lock-keeper's house on the other side, by the lock. It was just a small house, one room and another room upstairs, that's all it was. After Theo McCord, Matt Irvine lived in the lock-keeper's cottage and kept on the shop.

Wilmont? The Reades are out of it years ago and it was an old people's home. Is it still an old people's home? Oh, it's empty, is it? It was a beautiful house. Beautiful grounds. I remember when the Miss Reades lived there. They were tall ladies. I don't know their names at all, I just knew them as the Miss Reades. They had carriages. Open carriages. And they used to ride in the carriages and they always had a wee dog in the carriage. And they had a Dalmatian dog that ran alongside the carriages. But they always had these wee dogs. Long dresses and big hats with fancy hatpins.

No, I never was in the Wilmont House, I never was. Later the Dixon family lived there, Captain Dixon and his family. And they kept a staff of servants. There was Ballydrain Castle on the other side of the road. I didn't know the Miss Montgomerys. There was Hugh Montgomery, and I used to just hear my father talking about him. The last one was Miss Nellie, and then it was sold. Morrison lived in it for a while. The Morrisons, they took it. Mr Morrison, he was a big man and they had a business. They had two sons, Maynard and Clifford. And Doreen was the daughter. They had three children. They kept horses, you would meet them out early in the morning riding the horses. It was all horses around here years ago. Lovely horses.[123]

I never saw the ghost, I just heard about the ghost. There was one part of the Lagan no one would walk along at night because of something. There was really something. I remember hearing about it. On down from the Drum Bridge, on right round that corner, the lightermen even wouldn't be there, they wouldn't even leave their horses there no matter how late they were. It was towards Ballydrain. Then there was a lady that walked in the hollow of the lake, from Ballydrain down into the middle of the road. There was somebody there. But they used to play tricks on people. My Uncle Harry, he was always playing tricks. This man – I think it was a curate – and he was coming walking along and they lay on the top of the wall and they lifted his hat off. He took to his heels! My Uncle Harry and a neighbour, Jack Doran. They were always playing tricks! They wouldn't even leave their horses there. And I remember my grandfather talking about the ghost of Haddock.[124]

Mrs Higgin lived in Malone House. She was a tall, stout lady who rode a bicycle. There was a hill down past our house and she used to come whizzing down the hill. I didn't know her, I didn't know the people in Malone House. Later the Barnetts lived there after Mrs Higgin left. I think there was only Mr Barnett and his wife and a housekeeper. I didn't know anything about them at all.

After I left school I was at home quite a while. We had our own garden and we grew our own vegetables and potatoes and we had a meadow and I helped whenever he cut the hay. I always had to work.

My brother Percy, he moved off. My sister moved off, she was always busy. My brother John, he moved off. I was left to help my dad.

I was married, yes, 26 years I was married and lived in Station Road, Lambeg. My husband had a motor business up in Lisburn. I used to sit pillion on his motorbike. He was a good driver. We used to go to the seaside, we used to swim. We used to go to Helen's Bay, some near place. Sometimes on a Sunday we used to go to the Antrim coast.

Mr Brown, they had a house in Newcastle and used to go down there and I used to go too. With my father and my mother. My mother was cook to John Brown, my father and my mother were married then. Oh, a very, very elaborate cook. She just learned herself and she was a wonderful cook. Anything she made tasted good, even bacon and egg was lovely. It was different from anything. She was a born cook. She was the seventh sister, her father had seven girls. And they were all tall, all over six foot.

His mother lived on the Edenderry side, his mother Mrs Brown. She lived on the other side and he used to go across there. And there was a boat and you could have manoeuvred the boat, whatever way, I don't know what it was, I didn't know anything about it, but he could pull the boat. The boat was there, and he could get himself across and he used to visit his mother every night. She lived across the Lagan.[125]

We used to go to the Giant's Ring quite often, it was quite a walk down the fields. It was lovely. They used to have big sales of work and things like that. I remember they had a big thing at the Giant's Ring that lasted a week. There were all sorts that you could think of. There were shops and there was music and photographs, you could have got your photograph taken. It was a big festival. I would have been about 14, 13 or 14. A long time ago.

Down from Longhurst was Lismoyne. It belonged to the Calwells. I remember that house, yes. When it was sold, the people that bought it, Mr and Mrs Kinnaird, they had a family. There was Marjorie, Elsie, Lita, Jean, Helen and Nora and Jack. Two boys, Jack and one other. We used to all play together. The Brown children weren't there at that time because after Mr Brown died Mrs Brown left Longhurst. His son Eric joined the navy. I hadn't seen Eric for some time and Eric came to see me one day. And we were very glad to see each other. He was a tall stalwart boy, very good looking with lovely ginger sort of hair. No, brown, brown-ginger hair. Very handsome.

Lismoyne. There was a builder bought it and pulled it down. And it was ridiculous. You know, it shouldn't have been pulled down. I remember people used to come to my grandfather and ask him about Lismoyne. Nobody knew when it was built because Lismoyne was very, very old, but nobody knew when it was built. It was a huge house with very large rooms and there was a big ballroom in it and big fire

Right: Guests crossing on the Lagan ferry from Edenderry to Longhurst for a garden party given by Mr John Brown for the British Association, 1902.

Early portrait photograph of Auntie Girlie.

grates. The fireplace was all, oh, it was beautiful marble, and after everybody went, people came and took the marble out of it, people came and stole the fireplaces. Huge big grates.

My great grandfather McClinton was born at the Honeycomb, a place near the Lagan opposite Edenderry.[126] He had land and John Brown bought it to build Longhurst, a big house, a lovely big house. John Brown, I don't remember much about him – before my day, long before my day. My mother was from Ballygowan. She got fed up at home, that's why she came to Belfast. She was away to a clergyman's house, I don't remember his name, she was with him, looking after his children, his three children, his wife was a delicate woman. And then she came to Longhurst. I don't remember much about it. But she told me, it's only hearsay. The drawing room had a fire at each end. But then there was a big fire in the hall. A big fire in the hall. It was very high, it was built up. It was all stone. It had to be whitened every morning. It was electric light. John Brown had electric light. The electric was made. In the yard there was a building and, you know, you remember you used to have radio with the wee batteries? Well, that was how the light was run. The light was run off batteries. It was in a tall place. And batteries all in this. My father had to look after those. He seen that the batteries were all kept filled, you see. I don't know what they filled it with. The batteries were all kept filled and you could have heard the hissing! They used to hiss. John Brown had electric lights. And even the water was pumped up by electric. Everything was done by electric. Even had an electric iron.

I had a badger, Teddy. Kenneth brought him in here one night, a tiny wee thing of three months old and I took him. He was shivering and I warmed him milk, and Sidney McAvoy gave me a teat and I washed a bottle out and warmed him milk and gave him milk, and he had a wee box and I warmed a blanket and put it over him. He slept all night. I had him then 'til he was four years old. And Teddy got a virus and Teddy died. It was five o'clock and I was sitting up with him, it was an awful stormy night, oh it was a terrible night, and Teddy lay in my arms 'til he died. And he didn't like men. He liked women all right. It was men that had disturbed his home, you see. I had Teddy four years. Teddy used to follow me about just like a wee dog. I just went out and called him. Teddy, come on Teddy. I love animals. Listen to my heart. And dogs. The last three dogs I had here were Bouncer, a collie dog, I got him as a wee pup three months old. Then a friend gave me a wee Yorkshire terrier, but he wasn't a small terrier, he was a big fellow, Benji. And Dusky. I used to take Bouncer and Benji and Dusky and I used to go down the field every morning and we used to have great fun with a stick, we used to have great fun.

Roses go right back to when my sister was born
Eugene Trainor

I grew up in a place called the Cor Bog, my father worked the bog, and then we all grew up in it. It's just between Tynan Abbey Estate, Caledon and Castle Leslie Estate. There was about 10 farmers in Cor, each owning anything from between 10 acres and 50 acres, we were one of the 10-acre ones. So we just had a little smallholding, owned by ourselves. My dad worked at Tynan Abbey for 44 years. He worked on the agricultural side of things for about 35 years; in the last 10, give or take a year or two, he worked the gardens, a one-acre south-facing walled garden. And in the summertime, Twelfth of July and in August, when they would all be off parading, we would love it. Twelfth

Eugene Trainor amongst the roses, 2011.

and Thirteenth of July we would go over to the Abbey because it was the tomato season! In August, that was the grape season and we would help ourselves to the grapes in the greenhouses! They had oranges, grapes, lemons, even pineapples growing in there. Sir Norman Stronge, when he went away off or when he was over to Westminster, he would come back with some unusual plant to give dad to plant in the gardens. We helped close and open the greenhouses and then go over and weed and pull raspberries and that's how we became interested in horticulture, so there was only one way forward for me and that was horticulture. Now in those days I had no idea what horticulture was – it was gardening, you know. So I went to the usual schools and then went to Greenmount Agricultural and Horticultural College in 1968. That was the beginning, really, of my career: Horticulture at Greenmount, Askham Bryan in England and the Royal Botanic Garden in Edinburgh.

I started working with Belfast City Council in 1992, as parks superintendent in the Malone area, and at Sir Thomas and Lady Dixon Park in 1994, when the then head of parks asked me to go out and look after the international rose collection. The rose gardens were originally laid out by 1965 and the first rose trials, the first judging, was carried out in 1965/66.[127] The Rose Society of Northern Ireland was formed in 1964. And it was a very, very lucky break for me because I had always been interested in roses and when I look back over my career, roses go right back to when my sister was born. We were plucking roses when my sister was born, I was holding one in my hand. And when I worked in Edinburgh we planted a lot of Edinburgh City Bypass with roses, the roundabouts, et cetera, and the Scottish Widow's Headquarters roof gardens, we introduced roses there. So it was in my blood.

Rose breeders all around the world specialising in the production of new roses, hybrid teas and floribundas, now generally known as the large-flowered and cluster-flowered roses, want to bring new

Lion Dance, Rose Week, Sir Thomas and Lady Dixon Park, 2011.

roses to the attention of the public. This has been achieved by setting up trial grounds throughout the world, where visitors can go and look at the roses, feel, smell and touch the roses. We have trial grounds where the name of the breeder is put on it with country of origin and year of the trial. Any rose in the trial grounds – and this is really specific to Belfast – has to be available to the general public, available to anybody to plant in their own garden, parks, roadsides, wherever. And if the rose is doing well in the trial, then it will do well in the local area. They will do quite well in a wide range of soils, particularly in high clays where other plants will fail, and being deep-rooted they will survive the droughts. I mean, roses just do not like wet weather, they just love the sun.

The World Federation of Rose Societies got together a number of years ago and they have conventions where they award prizes for rose gardens. Not just the case of going around looking at rose trials, but to look for fantastically well laid out rose gardens, fantastic collections of roses. And they decided that they

Right: The rose gardens at Sir Thomas and Lady Dixon Park, 2010.

would award a Plaque of Merit for rose gardens every three years, at a convention. I had the privilege of going and representing Sir Thomas and Lady Dixon Park to the World Federation of Rose Societies Convention 2000 in Houston, Texas. I went to give the five-minute presentation. And they looked at all the other presentations, did their judging and they gave the award to STLD! We were the fourth ever recipient of the Plaque of Merit. There was happiness and merriment! I could not wait to phone up Maurice Parkinson, the head of parks. Four o'clock American time to hit nine o'clock British time. 'We won!'

The rose garden has gone from success to success. It's all down to the team. Through the hard work of the park's marketing and events officer, we have Rose Week in STLD, Sir Thomas and Lady Dixon Park. Our Rose Week is renowned throughout the world, not all trial grounds have a rose festival, and in Belfast a whole week in mid-July is dedicated to roses. We have 800 different varieties and get 30 to 40 or so new varieties every year and about 30 or 40 are taken out. There are about 45,000 rose bushes in the rose gardens all together. The roses are judged by the Rose Society of Northern Ireland, 20 to 30 judges, five times per season over three years. Marks are collated and the awards made. Visiting judges, rosarians from throughout the world, take part in the final 20 percent of the marking.

A team of five staff and two apprentice gardeners look after about 15 acres dedicated to rose growing including lawns and other areas. People ask, 'When do you start preparing for Rose Week?' And basically the answer is – 'As soon as one Rose Week is finished!' It's ongoing all the time. People can visit any time of the year, it's open from dawn to dusk with the exception of Christmas. One of the things about roses is that they are in flower, depending on the variety, from April right through to November or December, though there is nothing from then to March, so what we are looking at is companion planting, I think it's the way ahead with roses.

The management of roses. You spend the whole year growing it up and then at the end of the year you cut it all down – there doesn't seem a lot of sense in that, but that's what we do! The main thing is to keep the plant young, healthy and vigorous and floriferous. They are plagued with disease, and that has to be kept at bay with chemicals, but the future is to grow roses with zero chemicals. There has been a lot of research done over the years breeding out disease in roses.

There is most interest in the shrub roses and climbing roses. The large-flowered roses, I would say, are on the decline; the cluster-flowered are on the increase. But the day when we can get away from having this tremendous amount of management – pruning, spraying – the sooner it happens, the better, though it will certainly not be in my time. I do like the old roses, highly perfumed, disease resistant, they will grow year after year without having to put secateurs or sprayer near them. The OGRs, Old Garden Roses, these roses are virtually maintenance free. I have a particular passion for the OGRs, but I also love many of the new varieties, especially David Austin shrub roses.

I loved my childhood here in Drumbeg
Clara Crookshanks

My parents lived about 100 yards from where I live now, on the Quarterlands in Drumbeg. I was one of two children but there was too much of an age gap, we were never very close growing up because she was so much older. There were horses in the fields here, there was a couple of old Clydesdale horses which belonged to Sammy Dunlop who was a local farmer, and my father used to take them back to the farm at night and he would have put me on their back, you know, and taken me up to the farm. There was a brown and white one and the other was a darker colour. And you fed them with a potato. Out in the morning before going to school, just cut the potato in two and held it to them. The summers seemed to be long, you had the whole summer to run about and play in the fields. And we would have wandered all over the fields with the other children, local children. We never had any fear in those days. And then we used to go to the dye works dam and Lagan and fish, fish for wee spricks as we called them. There was a path from here up through the rural cottages and then down round through Gray's farm and round the dam at the dye works. The dye works was on the Drumbeg Road a short distance from the corner with the Ballyskeagh Road. It was sort of a shortcut to the bus stop. The rural cottages, there were 12 of them, they were like council houses. See there were few houses here, if you didn't work on a farm you couldn't get a house. After the dye works closed they filled in the dam and built the houses called the Hermitage.

My mother was born here and my grandmother was probably born here too. But my grandfather, he came from the lough shore, away out at Crumlin. He came to work here as a farm labourer and then he met my grandmother. But my father came from Shaw's Bridge, his father came up there to be a land steward on a farm at the Giant's Ring. My father he laboured for a while in Ballydrain and then he got a job in Marshall's meal company in Victoria Street. He worked there as a storeman. He worked there for 25 years, I am sure, but he was a farm labourer before that.

It was a really happy place to grow up. We had a great childhood, I loved my childhood here in Drumbeg. We had just one local shop, Mrs Gardiner and her two daughters, I still keep in touch with them. It was where Drum Cottages are now, on the Ballyskeagh Road. Then there was the shoemaker and Bob Stewart's pub. Of course it was a man's pub, women didn't frequent the pub. It would have been unheard of for a woman to go into a pub. The shoemaker was opposite the shop, almost opposite the shop. But I can remember going to the pub – my daddy kept pigs, you see, and when there was baby pigs, sometimes the sow needed Guinness to give it strength! And I used to go on the bicycle with my bag to the pub but I wasn't allowed to go to the pub door, you had to go to Mrs Stewart's front door. And she brought you in and then she would have went and got the bottles of Guinness. And she had a big stuffed bird. Always I can remember that stuffed bird! An owl. The pub was just one room, really. Mrs Stewart was a tiny wee woman. Bob was always an old man to me, he was always stooped over and short and he had a wee moustache, if I remember right. And he just slipped about, he didn't lift his feet.

Thompson's dye works. The works shut in the 1960s and the site was developed for housing.

The shoe shop was a dwelling and the shoe shop was one side. Two doors, one into the house, one went into the shoemaker's shop. An old wooden staircase going up, and he had all the machinery there and he had his stool and Jimmy would have sat on his stool. There was a window, he could have watched everything that was going on. Jimmy Stewart, we called him. The name was there, Jimmy Stewart, on the door, wasn't it? I don't think they were related to the Stewarts of the pub. He had a machine there for sewing round the shoes. They were men's shoes. Leather. Dress shoes he made. They would have been expensive. It would have gone on until he died around the early '60s. He actually owned the field opposite the Orange hall and he was very possessive of the field. And the boys, they knew that he was good for a chase, you know. So they would have went through the hole in the hedge and he would have been out and down and chased them away. He watched everything. He preached a bit, he would have got up and sort of given his testimony. There was always a mission, a little mission held in the Orange hall on a Sunday night. People that didn't go to church, I suppose, would have gone to the hall. He would have been singing at the top of his voice, you could have heard him outside.

Right: Clara and her older sister Margaret in front of the Irvine's lock house and shop at the sixth lock, Drum Bridge, in about 1946.

When we were children the Twelfth of July was like our holiday, that was the day you got off and you had a great day out and then you had a party at night, a lovely sit-down meal for the men, then a nice sing-song at night. Everybody had a song, you know. The singers had their own song and I can remember, 'The Lord Mayor had a coachman', by Sandy Gray, he sang it and Ricky Craig played the spoons and he would sing, 'The little old mud cabin on the hill'. Mrs Alexander, she would sing, 'Silver threads among the gold, darling I am growing old!' Aunt Maggie didn't sing much in my time. My daddy would sing too, but he wasn't a great singer, he would do, 'My granny's old armchair'. And you know, different ones were asked to sing, and there was another man called Baker and he used to get up to sing and then he repeated, they couldn't get him sat down. He wanted to sing on forever.

I remember a fridge coming to the shop, and that was a really big occasion when Mrs Gardiner got a fridge because we got ice lollipops! I mean, you only got ice cream when you were in Belfast or in Lisburn and that wasn't every week, you know. To have one on the doorstep it was great. They sold everything. Groceries, vegetables, sweets, washing powder, cigarettes. Gardiner's shop sold everything.

At the canal, the little lock house was Mrs Irvine's, it was just solely sweets and drinks, cigarettes. It was very, very tiny and you stepped down into it. Just a wee bit of a counter and the door opened right up against the counter. And Mrs Irvine was a nice wee white-haired lady and, ach, she was very pleasant to go into. And we used to have coupons after the war and you had to have coupons and your ration book with you. And you saved those up. There were big Gobstoppers, I remember, and drinks. Matt Irvine was the lock-keeper.[128]

But Drumbeg was great, we used to do Queen of the May and things like that. The children all got together and we went round the doors collecting money for the Queen of the May. We dressed up. There was always one who was dark, 'the darkie said he would marry me' – not politically correct now. That would have been a girl, the boys didn't really participate, it was mostly girls, and they would have sort of blackened her face, it was probably soot, but the queen – Joan Gardiner was always the queen. She was very pretty. She sat on the pram, we pushed her in the pram and we dressed her up like a bride, really. And then you went round all the doors and collected money and then when May Day came we went into Mrs Gardiner's shop and bought buns and we would have had a picnic. And we just spent the whole money on buns and sweets and drinks. Spent the money. And the children just had a party in the field.

My father and mother belonged to Ballycairn Church which is away round Ballylesson Road, and daddy used to take me on the bike, on the bar of his bike, and he rode the bicycle up Ballydrain Avenue and then he came down a little path onto the towpath and along there and then over the locks. And over two boards, the Eel Weir they called it. Just two planks of wood and you had to go over that on the bicycle and I used to be a bit afraid. And then through Edenderry and up to Minnowburn. His sister lived at Ballynahatty, there at the Giant's Ring, so we always ended up at Aunt Mary's.

I can remember them walking up between the yew trees
Matthew Neill

My father lost his job on the railway in 1932 – no early '33, and he was out of work, and for a couple of years we lived at Ballymacash and we were very poor, no money, no nothing. And we were down in the very dregs and my father started to help my grandfather down in Drumbeg. My grandfather was sexton of Drumbeg Church. In due course my father took over the job and then we came to live in Church House next to the Parochial Hall. I lived there until I was married, in fact. I was active in singing, always, and concerts, plays and so on. Great fun. And then, of course, I met my wife down here and when you are in love with a girl and you live in the same spot, well that's the place to be!

During the war the Americans were at Ballydrain for a while and the thing that I remember most about them is they had to have church parades. A lot of them would have come down to church, British and the Americans, we had to learn the American anthem and sing. They paraded down in uniform and round the church and in at the front. I remember well. And, of course, there was no room hardly for the congregation by the time you put 120, 130 servicemen in. The Americans were great fun because they were so lackadaisical, they came more or less sauntering down – all in step but, you know, a long way less military than the British.

Matt Neill at Drumbeg Parish Church, 2011.

I remember Sir Thomas and Lady Dixon, they came to church. He was a tall, pleasant man, she was a smaller person. I can't remember a lot about them, I was trying to think. I can remember them walking up between the yew trees.

When I retired in 1980, aged 60, I became the sexton here in Drumbeg. Doing the grounds, digging the graves. Cutting hedges, preparing for Sunday services. I dug graves by hand. Six foot six. Sometimes you would do it in a day, sometimes a day and a half. It's only six foot six by two foot six, you know, not big. Then a local man used to come in and help me, George Dougan. And other men came in from time to time too and helped out. Oh yes, oh yes. It gave me a great sense of continuity working there. And now Hugh is the sexton and Hugh is connected to the family because he married my niece.

Drum Bridge and St Patrick's Church.

I wrote the history of the church, the only history that ever was written.[129] Of course, yes, I would think of my father and grandfather working there before me. I remember one occasion somebody had died and they belonged to a family and they had two separate graves, one there and one there, and I thought, now, which one? And I looked at the names and the dates and I couldn't decide which one to open, where to put the coffin. But my grandfather wasn't there to ask. Now, if my grandfather would come along and tell me, I would know what to do, but he didn't, nor my father either!

This way leads us to some of the most important, well, best known, graves in Drumbeg. The most notorious grave is Haddock's grave which we come to now. It's noticeable because it's long since lying flat, or at least slightly sloped on the ground. It's said that long ago Haddock escaped and went back again, that's why it's like that. He appeared as a ghost. You can read, perhaps even yet, the text on the stone. 1657. It is said his hand appeared at a court of law. Of course the stones in all graveyards have a history behind them. My grandfather is buried over here, I must get a stone set up at his grave sometime! Now this is a celebrated stone, tours come to see this stone. He was the last of the volunteers of 1798, he died aged 104 it says. All of these have stories behind them. McCance of Dunmurry. Charley were another important family. Coates, they have quite a number of burials here. Montgomery. Look, Annabella Heron, 1909. It says, 'A faithful servant of Thomas Montgomery.' Montgomery owned Ballydrain at one time. Here is a stone worth looking at because it records a rather tragic accident that happened in Belfast. Seamen of the barque *Herzogin Cecilie*. She was, I think, a Scandinavian ship, one of the beautiful tall ships, and the boiler for the crane blew up and they were blown into the water dead. They had no burial place but they came to Drumbeg, that's where they were buried. 1935. Look, Belfastiensis. 'Isaac W. Ward, Belfastiensis, Chronicler of the history of his native city.' A nice title! 1834-1916. He died very tragically, he was struck by a pony and cart in North Street.

Those yew trees, my grandfather actually helped train those. They were provided by the people at Ballydrain. I think the date is on the plaque there. 1885. When my grandfather came here they were only young trees and he took over the training of them. They were only small trees, beginning to grow. The church provided iron hoops and they trained them up over the years. The bottom two are new ones, do you remember them being planted? But they have been allowed to grow too wide and deep. They need to be cut really hard back, tight in round again. I had to do it a couple of times. I remember once asking my father. My father said, 'Ach well, I don't mind, yeah, cut them if you like.' And I started with a saw and a pair of loppers and when I had finished there was nothing, only wee things like that over the top. And there are some lovely big trees in the graveyard. Quite ancient, too. Lime, beech, oak. A big copper beech, it's been there many, many years. It was there, just like this, when I was a boy. Oh, a long time ago.

The yew arches at Drumbeg Church by Niall Timmins, 1998.

Notes and References

1   Anon, *The Ancient and Present State of the County of Down* (Dublin: A. Reilly, 1744), p. 129; In 1755 Thomas Pottinger of Ballymacarret referred to the ferry which he said had operated at the time of his grandfather, see advertisement in the *Belfast News-Letter*, 6 May 1755, p. 2

2   Copy of an inquisition taken at Carrickfergus before the Bishop of Dromore, Sir Foulke Conway and Stephen Allen Esq. Public Record Office Northern Ireland (PRONI) D2977/5/1/1/7. This typed document is described as a 'copy of a copy' and is dated 1630, though the catalogue entry gives a date of 1621, which is presumably when the original manuscript was written.

3   Enrolment of Settlement. Exemplification of Grant. Various manors in County Antrim, Lough Neagh and the Bann. 20 July 1670. PRONI D389/9. This document refers to '...that Piscary or Fishing in and upon the River of Lagan neere Strandmellis...'

4   Advertisement by Thomas Pottinger in the *Belfast News-Letter*, 6 May 1755, p. 2. The claim to part of the river made by Pottinger was disputed by the Earl of Donegall, see advertisement in *Belfast News-Letter*, 30 May 1755, p. 4. No references to commercial fishing in the lower Lagan during the nineteenth century have been found and an intriguing note that a six foot five inch sturgeon was caught when 'Lord Belfast and some Gentlemen were drawing their nets at Ormeau' (*Belfast News-Letter*, 28 July 1815, p. 2) is most likely a fireside fishing tale. By the end of the nineteenth century a project to restock the Lagan with salmon and construct a fish pass at the first weir was being considered, see note in *Belfast News-Letter*, 14 July 1899, p. 5.

5   Advertisements in the *Belfast News-Letter*, 13 April 1764, p. 2; 7-10 October 1788, p. 3.

6   May Blair, *Once Upon the Lagan: The Story of the Lagan Canal* (Belfast: Blackstaff Press, 1981), p. 111.

7   Correspondence in the *Belfast News-Letter*, 20 June 1899, p. 3.

8   Anon, 'Belfast Corporation gas works', *Belfast Evening Telegraph*, 19 February 1907, p. 6.

9   Anon, 'Belfast Amateur Band', *Belfast News-Letter*, 4 June 1833, p. 4; Anon, 'Musical treat', *Belfast News-Letter*, 2 August 1833, p. 4; Anon, 'Musical Aquatic Excursion', *Belfast News-Letter*, 11 August 1835, p. 4; Anon, 'Musical Aquatic Excursion', *Belfast News-Letter*, 1 July 1836, p. 2. A different view of the 1836 excursion was given by *The Northern Whig*, 30 June 1836 p. 2, which noted that the event was spoilt when 'a number of the lowest rabble made their appearance' and shouted and threw stones. A small illustration of the Battery at Ormeau is given in Joseph Molloy, *Belfast Scenery in thirty views* 1832 (Belfast: Linenhall Library, 1983).

10   Anon, 'Arrival of Lord Hamilton F. Chichester – Rejoicings at Ormeau', *Belfast News-Letter*, 9 September 1842, p. 3.

11   Anon, 'Fete champetre, in the Botanic Garden', *The Northern Whig*, 11 August 1840, p. 2. See also 13 August 1840, p. 2.

12   Walter F. Mitchell, *Belfast Rowing Club 1880-1982* (Belfast: Belfast Rowing Club, 1994), p. 11.

13   Anon, 'A modern Robinson Crusoe', *Belfast News-Letter*, 2 January 1867, p. 2; Correspondence in *Belfast News-Letter*, 4 January 1867, p. 4; Anon, 'Canoeing on Lough Neagh', *Belfast News-Letter*, 14 January 1869, p. 4. It is highly likely that the man who paddled the canoe on the Lagan in 1867 was also one of the two men who took part in the adventure published in 1869, though unfortunately, the canoeists are not identified.

14   George Benn, *A History of the Town of Belfast* (Belfast: Marcus Ward and Company, 1877), pp. 557-558.

15   For an example of a particularly strongly-worded complaint about the state of the Blackstaff, see Anon, 'The Blackstaff nuisance', *Belfast News-Letter*, 22 September 1855, p. 2. The ballad appears in *The Irish Builder* (1874), p. 305. It is described as being 'slightly altered from some lines which appear in the *Manchester Lantern* on the Irwell River'.

16   Anon, 'The culverting of the Blackstaff', *Belfast News-Letter*, 22 April 1881, p. 5.

17   Correspondence in *Belfast News-Letter*, 13 October 1885, p. 6.

18   See for example, Anon, 'High-Level Sewerage Scheme. Special meeting of the Town Council', *Belfast News-Letter*, 27 October 1885, p. 5; Anon, 'The pollution of the Lagan', *Belfast News-Letter*, 11 November 1886, p. 8; Anon, 'The purification of the Lagan', *Belfast News-Letter*, 1 December

1886, p. 8; F. W. Lockwood, *Notes on the Sewerage of Belfast, and Pollution of the Lagan* (Belfast: 1886).

19 Anon, 'Review of general trade and commerce for 1890', *Belfast News-Letter*, 1 January 1891, p. 6.

20 An Act to empower the Lord Mayor, Aldermen and Citizens of the City of Belfast to construct a Lock and Weir on the River Lagan; to construct a Wharf Embankment and Roadways; to make Street Improvements... Belfast Corporation Act (Northern Ireland), 1924.

21 Anon, 'Big improvement scheme McConnell Lock and weir to be opened to-morrow', *Belfast Telegraph*, 29 November 1937, p. 11; Anon, M'Connell Lock and Weir opened', *Belfast News-Letter,* 1 December 1937, p. 6.

22 Robin Sweetnam and Cecil Nimmons, *Port of Belfast 1785-1985 an Historical Review* (Belfast: The Belfast Harbour Commissioners, 1985), p. 54. Although this reference states that coal deliveries by barge to the Gasworks ended in 1963, photographs of coal deliveries in the archive of NIEA Built Heritage are dated June 1964.

23 The poor condition of the lower reaches of the Lagan was highlighted in two reports: K. P. Bloomfield, *River Lagan Report of a Working Party* (Belfast: Department of the Environment NI, HMSO, 1978); K. E. G. Morrow, J. S. D. Gilmore and E. T. Morahan, *The River Lagan Stranmillis Weir to the sea. Report of the Northern Ireland Council for Physical Recreation and Sports Council for Northern Ireland Joint Working Party on the State of the River Lagan for Amenity and Recreational Use* (Belfast: 1979).
For details of the plans for revitalising the area, see Peter Hunter and Roy A. Adams, *Laganside* (Belfast: Nicholson and Bass Ltd, 1987).

24 For further details of the Belfast markets and the abattoir at around this time, see Anon, *The Belfast Book 1929. Local government in the city and county borough of Belfast* (Belfast: R. Carswell and Son, 1929), pp. 131-133. See also Anon, 'Market day in Belfast', *Belfast News-Letter*, 21 September 1929, p. 6.

25 Vaccines have now made whooping cough uncommon, but in the past, children were often taken to a gasworks to inhale the strong fumes to try to reduce the severe uncontrollable cough. Since recording this interview, reference to women in the Crumlin Road area of Belfast passing an infant three times under a donkey to affect a cure for whooping cough has been noted in *Ulster Saturday Night*, 19 January 1895, p. 1.

26 For further information about the development of the Gasworks, see Anon, 'Belfast Corporation Gas Works. The interesting history of the undertaking', *Belfast Evening Telegraph*, 19 February 1907, p. 6; Anon, '100 years ago and now Belfast gas centenary', *Belfast Telegraph*, 26 June 1923, p. 9.

27 During the early 1960s, the production of town gas at the Gasworks site ceased and gas was piped from a new plant in the harbour at Sydenham. Anon, 'New gasworks will use by-product of oil refinery', *Belfast Telegraph*, 11 May 1964, p. 2; Anon, 'Gas at the push of a button', *Belfast News-Letter*, 15 May 1968, p. 2; Anon, 'Another milestone in the history of Belfast Gas Department', *Belfast News-Letter*, 15 May 1968, p. 11.

28 Alan Whitsitt, 'Gas. Last days of an old flame', *Belfast Telegraph*, 27 May 1987, p. 9; Billy Simpson 'Saying goodbye to an old flame', *Belfast Telegraph*, 18 October 1988, p. 9. The Gasworks site was subsequently developed as a business park, though in the following decade, natural gas was piped from Scotland, and Phoenix Gas reintroduced a piped gas supply to houses in Belfast and other places.

29 For details about the history of St George's Market see the booklet Anon, *St. George's Market* (Belfast: Belfast City Council, [n.d.]). The relocation and closure of Belfast's markets started in the 1960s when it was proposed to move markets and the abattoir to the suburbs. One reason given was 'modernisation', another was to provide space for an inner ring road, though this was never built. See Anon, 'Belfast horse-and-cart age markets to be modernised', *Belfast Telegraph*, 15 May 1964, p. 5.

30 A booklet produced to celebrate the centenary of the Ormeau Bakery provides a summary of the business and its achievements. See Anon, *Ormeau Bakery 1875-1975* (Belfast: 1975). PRONI D2498/50.

31 For the history of the Botanic Gardens see Robert Scott, *A Breath of Fresh Air: The story of Belfast's Parks* (Belfast: Blackstaff Press, 2000).

32 'Ulster 71' was a festival aimed at promoting Northern Ireland to encourage investment, with a series of events and activities, many of which took place on the riverside fields by the Botanic Gardens and at the recently completed building that would become Queen's University Physical Education Centre. See Anon, 'Ulster 71. Festival with a serious purpose', *Ulster Commentary, 297* (1971).

33 Sightings of seals on the Lagan have become increasingly common. See Linda McKee, 'The Stranmillis seal', *Belfast Telegraph*, 24 April 2009, p 3.

34 There have been many articles about the improving condition of the tidal Lagan, see for example, Anon, 'River revival', *Ulster News Letter*, 5 February 1991, p. 17; Anon, 'Big business breathes life back into the River Lagan', *Irish News*, 24 May 1993, p. 5.

35 C. W. Russell and J. P. Prendergast, *Calendar of the state papers relating to Ireland in the reign of James 1. 1608-1610* (London: HMSO, 1874), pp. 88-90; Robert Pentland Mahaffy, *Calendar of the state papers relating to Ireland of the reign of Charles 1* (London: HMSO, 1901), p. 174.

36 Anon, *The ancient and present state of the County of Down* (Dublin: A. Reilly, 1744), p. 72.

37 Anon, 'Belfast', *Belfast News-Letter*, 10 September 1754, p. 2.

38 See *Journals of the House of Commons of Ireland*, second ed. (Dublin: Abraham Bradley, 1763), 11 1759-1760 pp. 344-391.

    For the history of the canal, see W. A. McCutcheon, *The Canals of the North of Ireland* (Dawlish: David & Charles, 1965); May Blair, *Once Upon the Lagan: The Story of the Lagan Canal* (Belfast: Blackstaff Press, 1981). See also Anon, 'Lagan lighters and lightermen', *Belfast Telegraph*, 10 August 1905, p. 3; J. H. Smith, 'When a spirits tax paid for water!', *Northern Whig*, 23 April 1954, p. 2.

39 M. G. P. Delany, *The Autobiography and Correspondence of Mary Granville* (London: R. Bentley, 1861), 3, p. 515. Arthur Hill was one of the signatories of a petition that proposed the construction of the canal and was involved in the project. See *Journals of the House of Commons of Ireland*, second ed. (Dublin: Abraham Bradley, 1763), 9 1751-1755 pp. 62-64; 11 1759-1760, pp. 344-391.

40 Anon, 'Belfast canal', *Belfast News-Letter*, 31 December 1793 – 3 January 1794, p. 3.

41 Advertisements in *Belfast News-Letter*, 16 July – 20 July 1779, p. 3; 2 October – 5 October 1787, p. 3.

42 Advertisement in *Belfast News-Letter*, 28 July 1797, p. 3.

43 Advertisement in *Belfast News-Letter*, 26 June 1829, p. 3.

44 Advertisement in *Belfast News-Letter*, 21 March 1855, p. 3 announced that 'Lockview House (the residence of the late Jas. M'Cleery, Esq.), situate near the first Lock' was available to let. Mr McCleery was Secretary to the Lagan Navigation; Francis Davis, *Rambles and Gossip along Highways and Bye-ways round Belfast* (Belfast: Alex Mayne, 1866), pp. 50, 53.

45 Anon, 'The Lagan Vale Estate Brick and Terra Cotta Works, Limited', *Belfast News-Letter,* 12 April 1897, p. 4; Anon, 'New issue. The Lagan Vale Brick and Terra Cotta Works, Limited', *Belfast News-Letter*, 12 April 1897, p. 8.

46 Advertisement to let mill and factory sites by canal placed by Lagan Vale Estate, *Belfast News-Letter*, 13 April 1897, p. 2; Anon, 'Vulcanite a thriving Irish industry', *The Irish Builder and Engineer,* 47, (1905), pp. 655-659. See also Fredrick Gilbert Watson, *Building over the Centuries. A History of McLaughlin & Harvey* (Belfast: Nicholson & Bass, 2010), pp. 71-72.

47 Leister's Dam was most likely named after John Leister, who held land in this area in the early nineteenth century. See plan of the Belfast water-course surveyed January 1837 in Edward Wakefield Pim, *Sketch of the Rise and Progress of the Water Supply to Belfast* (Belfast: W. G. Baird,1895).

48 For information about Belvoir Park, see Ben Simon, *A Treasured Landscape: The Heritage of Belvoir Park* (Belfast: Forest of Belfast, 2005); Newforge is described by Kathleen Rankin, *The Linen Houses of the Lagan Valley* (Belfast: Ulster Historical Foundation, 2002), pp. 200-202. See also the commentary in the reprint of Joseph Molloy, *Belfast Scenery in Thirty Views 1832* (Belfast: Linenhall Library, 1983).

49 Ben Simon, *A Treasured Landscape: The Heritage of Belvoir Park* (Belfast: Forest of Belfast, 2005).

50 There were forges and ironworks near the Lagan at Stranmillis, Lambeg, Newforge and Oldforge (Dunmurry). See Eileen McCracken, 'The woodlands of Ulster in the Early Seventeenth Century', *Ulster Journal of Archaeology*, 10 (1947), pp. 15-25.

51 Ben Simon, *If Trees Could Talk: The Story of Woodlands around Belfast* (Belfast: Forest of Belfast, 2009), pp. 22-23; Advertisement in the *Belfast News-Letter*, 14 October 1808, p. 1; E. R. R. Green, *The Lagan Valley 1800-50* (London: Faber and Faber Ltd, 1949).

52 Anon, 'The water rights of the Lagan Canal', *Belfast News-Letter*, 23 October 1884, p. 7; Anon, 'Belfast quarter sessions', *Belfast News-Letter*, 23 January 1885, p. 7.

53   The clog works at Newforge is listed in the *Belfast Street Directory* between 1918 and 1922.

54   Alfred S. Moore, 'Newforge perpetuates name of city's first foundry', *Belfast Telegraph*, 6 December 1951, p. 4. In 1964 Spillers acquired Newforge factory, though the Wilson family retained the surrounding land which later became a park. Information kindly provided by J. W. Wilson OBE in 2006. See also Anon, 'Pet food firm sold for £7m', *Belfast Telegraph*, 8 May 1964, p. 20.

55   Correspondence about bathing in the Lagan at Stranmillis: *Belfast News-Letter*, 7 August 1875, p. 3; 16 June 1877, p. 4; *Belfast Evening Telegraph*, 21 June 1906, p. 2.

56   Cathal O'Byrne, *As I Roved Out: A Book of The North* (Belfast: Irish News, 1946), pp. 245-248.

57   Francis Davis, *Rambles and Gossip along Highways and Bye-ways round Belfast* (Belfast: Alex Mayne, 1866).

58   The original building, a small rectangular structure, is shown, unnamed, on the first edition six-inch Ordnance Survey map of 1833. The building is also shown on a map in the book by Edward Wakefield Pim, *Sketch of the Rise and Progress of the Water Supply to Belfast* (Belfast: W. G. Baird, 1895). However, this map, a plan of the Belfast water-course surveyed January 1837, assigns the land by the cottage not to Molly Ward or to John Ward but instead to a William Ward. Further research is needed to unravel the early history of this building and the Ward family.

The later history of the building is revealed in the first revision Ordnance Survey six-inch map of 1858, which shows an enlarged structure named 'Molly Wards' with riverside gardens and summer houses. The Ordnance Survey map of the Borough of Belfast published at a scale of five-foot, sheet 66 of 1894, shows that this enlarged structure comprised two rectangular buildings connected together (most likely the original building and an annex). An advertisement for the sale of lands at Lagan Vale, Stranmillis, in 1855 (*Belfast News-Letter*, 17 August 1855, p. 3) included 'the Lagan Tavern, not many years built, licensed, and doing a fair trade.' This was without doubt Molly Ward's, the comment about it having been recently built was probably a reference to the construction of the annex. A later advertisement for the sale of Lagan Vale farm (*Belfast News-Letter*, 14 September 1859, p. 2) included 'a large Cottage known as Ward's Tavern, Licensed, and doing a good business.'

59   Anon, 'Easter Monday in Belfast', *Belfast News-Letter*, 14 April 1857, p. 3.

60   The earliest reference found to Ellen Keely (sometimes Keeley) was in 1879, when she was granted a new license at 'Molly Ward, Lower Malone', see *Belfast News-Letter*, 9 July 1879, p. 7. In 1882, she was advertising to sell the public house 'formerly known as Molly Wards', see *Belfast News-Letter*, 21 April 1882, p. 1. According to notes published in the *Belfast News-Letter*, 13 January 1936, p. 6 and 15 January 1936, p. 6, the McAuleys subsequently took over the tavern but it later became disused.

61   Anon, 'Ligoniel Petty Sessions', *Belfast News-Letter*, 28 September 1882, p. 6. A brief note about Molly Ward and the tavern published in *Ulster Saturday Night*, 13 April 1895, p. 1 seems to indicate that by this time both Molly and the tavern were only memories.

62   Information kindly provided by Dorothy McBride, daughter of George Kilpatrick, in 2011.

63   William Gratten and Richard Belshaw, lock-keepers and Jack Jones, bank ranger are mentioned in May Blair, *Once Upon the Lagan: The Story of the Lagan Canal* (Belfast: Blackstaff Press, 1981), p. 5. The first reference in the *Belfast Street Directory* to W. Gratten, lock-keeper is in 1918 (entry under Lockview Street).

64   In 1903, it was announced that 'F. King and Co., Ltd. makers of the well-known Edward's Desiccated Soup have decided to erect new works at Stranmillis, Belfast', *Irish Builder and Engineer* (1903), p. 1816. The factory is first listed in the *Belfast Street Directory* in 1905. It closed in 1960, though it was reported that four members of staff decided to form a company to continue to manufacture pickles and sauces. Today supermarkets still sell Edwards Pickles, though there does not appear to be any connection with the original business. Anon, 'Business to close down', *Belfast Telegraph*, 15 June 1960, p. 12; Anon, 'Powdered soup firm in liquidation', *Belfast Telegraph*, 2 December 1960, p. 5.

65   Michael Taylor was a well-known character, his death was reported at length in the *Belfast Telegraph*, 3 February 1925, p. 8. He had been born at the second lock in 1835 and died just before his ninetieth birthday. He had served in the army in India, Egypt, China, Ceylon and Africa, and after retiring from military service, for 40 years he was lock keeper at the house where he had been born. He later lived at 6 Wansbeck Street.

66   The glass bottle works was opened in 1899 by David Wright and Co. Ltd. It was quite close to the river and although large quantities of coal and glass sand were imported, road transport rather than lighters were used. *Belfast News-Letter*, 20 October 1899, p. 3. The factory is last listed in the *Belfast Street Directory* in 1924.

67   Mervyn explains why Laganvale Street, where some houses have a flat tarred roof, became known as Tar or Tower Row. Other local residents have suggested that perhaps Harleston Street became known as Piano Row because they had a front room with space for a piano. No one from the area could suggest why Wansbeck Street should have been known as Poverty Row.

68   Many of these people are listed in the *Belfast Street Directory*. For example, the directory for 1930 lists in Laganvale Street: Jas. Ingram, bottlemaker, at number 29; Thos. Slinn, bottlemaker, at number 35; and Mrs Wright at number 43. An article about the Stranmillis glassworks (*Belfast News-Letter*, 20 October 1899, p. 3), which noted that there had been a difficulty in getting experienced workmen and that 'nearly all the adult labour at present employed has been brought from England and Scotland', supports the comment made by Mervyn that the glassworks employed people from Pilkington glassworks in England.

69   Again the *Belfast Street Directory* confirms many of the details given by Mervyn. For example, the directory for 1930 lists John Callaghan, grocer, as living at number 1 Wansbeck Street. In 1935, this building was occupied by John Nesbitt, grocer. The directory for 1930 also lists W. H. Houston, grocer occupying a property on Lockview Road at the junction with Wansbeck Street.

70   Information about some of the factories in Lockview Road are given in references 64 (the soupworks), 73 (the brickworks) and 76 (Vulcanite factory). Another long-established business in this street was Campbell, Brown and Co. Engineers. The company is recalled by local people as having undertaken maintenance work on machinery in factories including Vulcanite and Ormeau Bakery and also metal fabrication such as making railings.

Dennis Curry, a chemist at Irish Cold Bitumen on the Stranmillis Road from 1952 to 1977, kindly provided the following information about this factory: 'They used to spray bitumen on roads followed by a squad of men that put on stones and then by a roller. It was usually for minor

roads but they did part of the M1 Motorway. It later became ICB Ltd. and then Shell took it over and it moved to Portadown. They also sold building products under the name of ''Colas'' and in the same place was a paint factory, called Montgomerie Stobo, where they made paint and tins for the paint, it was called ''Ulster Tin''. The factory buildings ran down behind the houses on Ridgeway Street to the Lyric Theatre and out by an entry at the theatre.'

71   In this area there were small brickfields shown on the first edition six-inch map of 1836. However, the third edition six-inch map of 1902 shows the riverbank between Ormeau and Annadale lined with brickworks. These were named (south to north) the Annadale, Prospect, Marquis, Ulster, Ormeau and Haypark Brickworks. The Annadale Brick Company Ltd. first appears as lessee in the 1888 Valuation book and brick making on an industrial scale appears to have begun at this site in this decade. To the north the Prospect Brickworks of H. and J. Martin commenced operations in the 1870s. The last to cease operations was the Prospect Brickworks, where a working face was observed in 1952. See P. I. Manning, J. A. Robbie and H. E. Wilson, *Geology of Belfast and the Lagan Valley* (Belfast: HMSO, 1970), p. 164. Further information kindly provided by Stephen Gilmore of Northern Archaeological Consultancy Ltd.

72   A short article written in 1958 briefly describes some of the places recalled by Ernie Andrews, noting 'the hulks of some of the old barges mouldering away on the Lagan river side of the locks but in the canal itself, beyond the lock gates, there is still a strange assortment of craft' and 'that abomination of abominations, the tiphead established on the canal bank by Belfast Corporation between the wars. It can stink still.' See, 'The Roamer', 'Out and about', *Belfast Newsletter*, 10 April 1958, p. 4.

The building Ernie described as 'Gibson's house' was also known as Newforge House. A photograph of this building around the time when Ernie played in it, windowless and with no front door, is given by Kathleen Rankin, *The Linen Houses of the Lagan Valley* (Belfast: Ulster Historical Foundation, 2002), p. 201. Ernie Andrews has described elsewhere his childhood visits to Belvoir Park. See Ben Simon, *A Treasured Landscape: The Heritage of Belvoir Park* (Belfast: Forest of Belfast, 2005), pp. 57-59.

73   The Lagan Vale Brick and Terra Cotta Works Ltd. operated from the late 1890s and closed in 1958. See Anon, 'New

Issue', *Belfast News-Letter*, 12 April 1897, p. 8; P. I. Manning, J. A. Robbie and H. E. Wilson, *Geology of Belfast and the Lagan Valley* (Belfast: HMSO, 1970), p. 164. The brickworks are described in Anon, 'The Belfast Mechanical and Engineering Association', *Belfast News-Letter*, 9 May 1899, p. 6; Anon, 'Brick and Terra Cotta Works, The Lagan Vale Estate, Belfast,' *The Irish Builder and Engineer* (1928), pp. 862-863. The manufacturing process is described in Anon, 'Laganvale Brick and Terra Cotta Works', *Belfast Naturalists' Field Club Proceedings*, 10 (1939-1940), pp. 41-42. A sketch of the layout of the factory is given on the cover of a catalogue produced by the brickworks, see Anon, *Lagan Vale Estate Brick and Terra Cotta Works Ltd* (1909). A copy of this rare publication is available in the National Library of Ireland.

74 Although 'Edwards Pickles' are still sold, the company that produced Edwards' Soup appears to have gone. See reference 64.

75 The soup works is described in reference 64. The red brick building with concrete window surrounds and clock tower on the corner of Lockview Road was designed in 1950 for the soup works. It was later occupied by Lamont, the weaving company, and included a factory shop. It is now offices. Paul Larmour, *The Architectural Heritage of Malone and Stranmillis* (Belfast: Ulster Architectural Heritage Society, 1991), p. 160.

76 See Anon, 'Vulcanite: a thriving Irish industry', *The Irish Builder and Engineer*, 47 (1905), pp. 655-659; Paul Larmour, *The Architectural Heritage of Malone and Stranmillis* (Belfast: Ulster Architectural Heritage, 1991), pp. 88-89.

77 For the history of Belvoir Park including the mansion house and gardens, see Ben Simon, *A Treasured Landscape: The Heritage of Belvoir Park* (Belfast, Forest of Belfast, 2005).

78 The *Belfast Street Directory* first lists R. M'Curley, lock-keeper (entries under Lockview Road) in 1915. Michael Taylor, the previous lock-keeper, was also recalled by Albert Allen in his stories of Stranmillis. See reference 65.

79 A well 'to the south of the churchyard, and close to the banks of the stream' was noted by James O'Laverty, *An Historical Account of the Diocese of Down and Connor*, 2 (Dublin: M. H. Gill and Son, 1880), pp. 222-223.

80 Albert Allen grew up in Laganvale and the Mr Brown he referred to was probably the Wm. Brown, shepherd, noted

in the *Belfast Street Directory* for 1930 as living at 18 Wansbeck Street.

81 The sawmill was situated close to the weir at the southern end of Morelands Meadow. See Ben Simon, *A Treasured Landscape: The Heritage of Belvoir Park* (Belfast, Forest of Belfast, 2005), p. 31 and maps on pp. 28, 44.

82 In mid March 1937, there was extensive flooding in Northern Ireland after heavy rainfall and the melting of a thick layer of snow. The Lagan flooded the second and third lock-keepers' cottages to a depth of two to three feet, forcing the occupants upstairs, where they were trapped. The Kilpatrick family were reported to have been fortunate to have had a stock of food in their shop. At the second lock Peter Rowan left to get food, but on his return could not reach the house because of the flooding, and a rope line had to be thrown across the water to take a basket of food to his wife and two children. At the first lock a section of the bank just above the lock was washed away and a number of barges were swept downstream. See *Belfast News-Letter*, 19 March 1937, p. 7, 20 March 1937, p. 7; *Northern Whig and Belfast Post*, 19 March 1937, p. 7, 20 March 1937, p. 7.

83 May Blair in her lovely book about the Lagan Canal provides the following details about the Kilpatrick family and their association with the canal: 'The last lock-keeper [at the third lock] was George Kilpatrick, who arrived there in 1922 and stayed until his death in 1968. The Kilpatrick family tradition of working on the canal was similar to that of many canal folk. George was born in 1891, the sixth of seven children of Alexander Kilpatrick (1849-1941). Alexander was a lock-keeper as was his twin brother Johnny (1849-1922). The brothers married sisters Anne and Liza Lynch whose brother George became manager of the canal in 1909. All three of Alexander's sons (James, Johnny and George) also became lock-keepers. When George started work at the age of fourteen, his first job was that of driving a horse and trap for Charlie Magowan, the then manager of the canal. Later he worked with the repair squad and eventually he also became a lock-keeper. He and his wife Sara reared ten children by the side of the Lagan and their youngest son Stanley still lives in the lock-house.' May Blair, *Once Upon the Lagan: The Story of the Lagan Canal* (Belfast: Blackstaff Press, 1981), p. 9.

84 The floods of mid March 1937 are described in reference 82. Another time when the Kilpatrick's lock house was

flooded was recorded in a letter from John S. Brown & Sons Ltd. to the Ministry of Commerce of 7 January 1957, that described recent flooding at the Edenderry factory and noted that 'There is an excellent lock keeper in the person of Geo. Kilpatrick at No. 3 Lock, and it was indeed pathetic to see the water surging round his house, about a foot deep, destroying his home and furniture.' PRONI AG5/6/9.

85   The factory is described in Anon, 'The Newforge story, turning a new page in industrial history', *Belfast Telegraph*, 29 October 1953, p. 4; R. G. NcCadden, *The Wilson Story* (Belfast: Nicholson and Bass, 1978).

86   In 2006, Mrs Helen Barbara McAslan, daughter of Clement Wilson, kindly gave the following story of how her father set up the factory at Newforge: 'My father, Robert Clement Wilson, had a head-on collision with his father when he was about 23 and his father said to him, "Well if you think you can run a business better than me, I will give you some capital and you go and prove it". It shocked daddy to the core. He had to leave his country, Scotland, his friends, and go somewhere, he didn't know where. He wandered around for a while, various places, and landed here and he saw this area at Newforge. He knew nobody here but he decided that this is where he wanted to be.'

Mrs McAslan also described her father's interest in landscaping. 'Daddy used to say he never saw a field, he always saw what he could do with it. He had this fabulous head gardener, Langley, and he would go round with him every morning and he would tell Langley what he saw in his head and Langley would do it. I always said it was his form of painting; he loved pictures but couldn't paint himself. Between the two of them they created not just the garden here but around all of the factories he had. Wherever he went, he always created a garden.'

87   Anon, 'Belfast's new industry. Corporation withhold sanction for factory', *Northern Whig and Belfast Post*, 2 July 1929, p. 9; Anon, 'Newforge factory resident director replies to opponents', *Northern Whig and Belfast Post*, 11 July 1929, p. 5.

88   Mervyn Patterson kindly added the following comments about the use of lighters by the Newforge factory in which he recalled two boats being used: 'If you go up the towpath past the second lock, you then reach the Red Bridge. A wooden structure, the buttresses are stone. A man called Wilson, the manager of Newforge factory, he decided that he would use the river to transport goods and

bought the two motorised barges. Well, each of them had a funnel and the funnel wouldn't go through the bridge there, it wasn't high enough. If you go up there now you will see that the bridge has been ramped up at each end. He raised the bridge so that the motorised barges could get through. One of the lighters was called *Eva* or *Ida* and the other had the name *Nellie*.' See also May Blair, *Once Upon the Lagan: The Story of the Lagan Canal* (Belfast: Blackstaff Press, 1981), p. 44, for a photograph of Newforge showing the motor lighter *Ida* 'bought by the factory at Newforge which used it to transport canned goods to the Belfast Docks'.

89   Evidence of some kind of early timber structure at this site was found when excavations for the new bridge revealed large oak timbers that were dated by dendrochronology to have been felled in 1617 and 1631. See David Brown and Mike Baillie, 'How old is that oak?' in Ben Simon, *A Treasured Landscape: The Heritage of Belvoir Park* (Belfast: The Forest of Belfast, 2005), pp. 85-97.

90   A crossing marked 'Shawes Bridge' is shown on William Petty's Downe Survey map of the barony of Belfast *c.* 1657, see PRONI T2313/1/17. The reference to the bridge being formerly small is found in Anon, *The Ancient and Present State of the County of Down* (Dublin: A. Reilly, 1744), p.129. Rebuilding is also suggested by comments made in 1809 by a local resident, Mr Legg, that there had been 'large stones and pieces of walls lying in the river which were removed when the Canal was making'. In addition, there is an old story that there was formerly a wooden bridge across the Lagan in this area. See the diary of John Templeton of Cranmore (entry for 1 September 1809) preserved in the archive of the Ulster Museum. An article by Colin J. Robb, 'Shaw's Bridge a glance at its history', *Belfast News-Letter*, 14 June 1955, p. 4, states that a Cromwellian document dated 1655 refers to the building of an early bridge by a 'Captt Shawe' though unfortunately, the source of his information for this (and many of his other newspaper articles) is not given.

91   Anon, 'The new bridge that forges a final link', *Belfast Telegraph*, 29 March 1977, pp. 6-7.

92   Anon, *Malone House* (Belfast: Ulster Architectural Heritage Society, 1983). In 2006, Clara Curry, who was born in 1918 in a cottage at the top of Dub Lane (her father and grandfather had both been land stewards of the Harberton Estate), kindly provided the following memories: 'I remember Mrs Barnett, sitting in a little summer house.

And I am sure she would have loved us to talk to her, she had no family and he, Mr Barnett, was a stiff old man, very stiff, very off-putting, hoity-toity. Grain merchants they were. He was, well, a fairly good-looking man. She was lovely but we never went near her, we just were sort of frightened to go, my brother and I. They had Violet the maid and Miss – oh, I can't remember her name, she was very friendly with us – Miss Cordiner. She was the housekeeper, she was a lovely, charming woman, very cultured. They would have had a cook and then Violet would have been the one who served the table, she was a very dainty looking person, from the west of Ireland. Mr Graham, his chauffeur, lived with his family in the yard, where they are making an outdoor sports centre. That's where the chauffeur lived, and he had four girls and two boys. Mr Barnett's car was a very big Minerva, it was a creamy coloured one, they had it for years in the garage. A very big grand car. And he had a big black one and something else. Far too much for one man!'

Clara Curry also related an intriguing local story that a tunnel used to run from the basement of Malone House in the direction towards Shaw's Bridge. It is not known if this story is true, though beyond the main basement of the mansion house there is an outer passageway around the house which was probably at one time open to the sky and provided light to the basement rooms.

93 The history of these houses has been described in a series of papers by Eileen Black: Eileen Black, 'Wilmont, Dunmurry: a profile', *Lisburn Historical Journal*, 4 (1982) pp. 37-46; Eileen Black, 'Ballydrain, Dunmurry – an estate through the ages', *Lisburn Historical Journal*, 5 (1984) pp. 17-28; Eileen Black, 'A glimpse of Drumbeg 1750-1800' *Lisburn Historical Journal*, 7 (1989) pp. 14-23; Eileen Black, 'Drumbeg 1800-1860' *Lisburn Historical Journal*, 8 (1991) pp.17-25. See also Robert Scott, *A Breath of Fresh Air: The Story of Belfast's Parks* (Belfast: Blackstaff Press, 2000).

94 Robert M. Young, *Belfast and the Province of Ulster in the 20th Century* (Brighton: W. T. Pike and Co, 1909), pp. 200, 351-352: Anon, 'Obituaries. Mr John Brown, F.R.S.', *Belfast News-Letter*, 2 November 1911, p. 8.

95 Anon, 'Rambles by road and by rail. No. LXIV County Down continued – Purdysburn', *Irish Farmers' Gazette*, 19 (1860) p. 627.

96 Anon, 'Treatment of mental cases. Modern methods at Purdysburn', *Belfast News-Letter*, 21 October 1929, p. 12. The recent history of the houses in Purdysburn Village was provided by local residents. For the history of Purdysburn Estate, see J. Fred Rankin, *The Heritage of Drumbo* (Drumbo: Parish of Drumbo, 1981).

97 Anon, *Extracts of Acts of Parliament, Relative to the Lagan Navigation, and Town of Belfast; also, of the Charter of the Town* (Belfast: Joseph Smyth, 1812), pp. 39-40; Advertisements in the *Belfast News-Letter*, 16 November 1756, p. 3; 2 June – 6 June 1775, p. 1; 14 July – 18 July 1794, p. 3.

98 Denis S. Macneice, '*Factory Workers' Housing in Counties Down and Armagh*', unpublished doctoral thesis, Queen's University Belfast, 1981; Additional information from residents.

99 The threatened destruction of the Giant's Ring is described in a note in the *Belfast News-Letter*, 8 December 1837, p. 2; Robert Young, 'The duty of preserving national monuments', *The Irish Builder*, 14 (1872), pp. 123, 140-143. Lord Dungannon's party is described in Anon, 'Novel and interesting soiree – Lord Dungannon and his tenantry', *Belfast News-Letter*, 3 August 1849, p. 2.

100 Advertisements in the *Belfast News-Letter*, 24 December 1858, p. 2; 28 December 1886, p. 8. Programme for the Grand Bazaar and Summer Fete at the Giant's Ring 1923, PRONI D3670/B/18.

101 'Drum Bridge and Towne' by a river crossing is shown on William Petty's Downe Survey map of the Barony of Belfast c. 1657, PRONI T2313/1/17. The story of a wooden footway that predated Drum Bridge is recorded in the diary of John Templeton of Cranmore (entry for 1 September 1809) preserved in the archive of the Ulster Museum.

102 Anon, 'Belfast', *Belfast News-Letter*, 9 September 1763, p. 2. No paper copy or good reproduction of this issue of the News-Letter is available in Belfast and the text was copied as accurately as possible from a poor quality microfilm.

103 Anon, 'Future of Lagan Canal', *Belfast News-Letter*, 3 July 1948, p. 3. Files concerning the abandonment of the Lagan Navigation include a letter dated 13 June 1956, which states that 'There is now no commercial traffic on the lower reaches of the canal and the Ministry has been contemplating taking early steps for the abandonment of that portion of the canal...' PRONI AG5/6/9. According to W. A. McCutcheon, in the spring of 1958, the Ministry announced its intention of making an order for the abandonment of the lower reaches of the canal from 1 July of that year. See W. A. McCutcheon, *The Canals of the North of Ireland* (Dawlish: David & Charles, 1965), p. 61.

104 The problem of fighting and drunkenness at the Goat pub is described in Anon, '"The Goat" application for new 7-day licence', *Belfast Evening Telegraph*, 18 January 1907, p. 3. After a number of attempts to have it shut, the pub was eventually bought not by the temperance movement but by a consortium of concerned local people who then let it as a grocer's shop. Alec Wilson '120 Co. Down magistrates sat in judgment on "The Goat"', *Belfast Telegraph*, 4 December 1950, p. 6.

105 Many of the people mentioned by Peter Johnston are listed in the *Belfast Street Directory* for Milltown. For example, the directory for 1947, p. 1499 refers to Alfred J. Downey, breadserver; S. Fairweather, grocer and hardware; Thomas Johnston (Peter's father); Hugh Owen, grocer; William McQuoid, farmer. See also Anon, 'Down your street', *The Ulster Star*, 14 April 1962, pp. 14-15, 22.

106 Alec Wilson, 'This village had a social service 50 years ago', *Belfast Telegraph*, 22 February 1951, p. 4.

107 Merrill Charlton kindly provided the following memories of Milltown: 'My father James Graham Gray had a dairy farm with 30 to 40 milking cows. Later in life he changed to pigs and built a modern piggery. The farmhouse was along a short avenue with beech trees, on the Milltown Hill. Later when the farm was sold, the house was demolished and the Barkley Restaurant was built. I have since heard that it is now apartments. The road called Gray's Park in Belvoir Estate was my father's field, bought by the council for housing and named in his honour.'

108 See Anon, 'Armagh and Shaw's Bridge are the latest targets', *Belfast Telegraph*, 22 July 1972, p. 1. A photograph of the Bailey bridge constructed over Shaw's Bridge is given in the *Belfast Telegraph*, 24 July 1972, p. 1.

109 Mrs Weir, who lives in Milltown (in a cottage built specifically for the residence of a district nurse), kindly provided the following local names of places near Milltown: 'There is a bridge over the Minnowburn, on the Ballylesson Road, just upstream from Minnowburn Beeches, called the Crooked Bridge. Sure enough, it has a bend on it. The lane at the back of the Ramada Hotel, from the dual carriageway to the Lagan, at one time extend to the Crooked Bridge. Its real name was Logwood Lane. The family Spence had a cottage on the lane, below what is now the Ramada Hotel. The lane was often called Kilpatrick's Lane, as they had the lock cottage. Round Purdysburn Demesne, by Hospital Road and going up the hill towards Purdysburn asylum, was a large stone wall; there are bits of it still standing. This was Dixon's Corner. If you were going that way you would say you were "going up the back of the wall" or "down the back of the wall". The old road from Shaw's Bridge to Malone Road was always Legg's Hill, Legg (sometimes Legge) owned Malone House at one time.'

110 The design of Purdysburn, with villas spaced out in attractive countryside, was innovative. See Anon, 'Belfast and district lunatic asylum. The villa colony scheme', *Belfast Evening Telegraph*, 7 August 1906, p. 6; Anon, 'Asylum or garden city', *Northern Whig*, 18 April 1910, p. 9.

111 For further information about the farm at Purdysburn see, Anon, *Belfast Mental Hospital* (Belfast: W. & G. Baird, 1924); Anon, *The Belfast Book 1929. Local Government in the City and County Borough of Belfast* (Belfast: R. Carswell and Son, 1929).

112 Purdysburn House was demolished in August 1965, though some associated buildings remain. See C. Douglas Deane, 'All that is left of Narcissus Batt', *The Newsletter*, 12 May 1979, p. 4.

113 Today the pond is drained and the roller skating rink has disappeared. The ice house remains, though hidden in a conifer plantation. The history of the roller skating rink is a bit of a mystery. There is little doubt that it was used for roller skating as it is described as such by Douglas Deane (see reference 112) and when both Peter Johnston who grew up in Milltown and William Glover who worked at Purdysburn since 1943 were asked about this concrete area, both referred to it as having been a roller skating rink. William, like Essie Murray, thought that it had predated the hospital and had been built by the Batt family when Purdysburn was a private estate.

114 Almost everyone spoken to about Edenderry had different local names for the three streets of terraced houses in the village! For example, one person who worked at Edenderry factory referred to the first row of housing in Edenderry as the Front Row, the street going up the left as New Row and the street ahead as the Old Row. Denis Macneice called the street at the entrance to the village and the street to the left both as New Row and the street straight ahead as Red Row. Yet more names for these streets are given in the stories told by Susie White, Annie McCarter, Marion McWilliams and Alan McWilliams.

For further information about the development of Edenderry and the factory, see Anon, *The Industries of Ireland, Part 1, Belfast and the Towns of the North* (London: Historical Publishing Company, 1891), p. 84; W. A. McCutcheon, *The Industrial Archaeology of Northern*

Ireland (Belfast: HMSO, 1980), p. 311; Denis S. Macneice, *'Factory Workers' Housing in Counties Down and Armagh'*, unpublished doctoral thesis, Queen's University Belfast,1981; Kathleen Rankin, *The Linen Houses of the Lagan Valley* (Belfast: Ulster Historical Foundation, 2002), pp. 194-199.

Mrs Weir, who has a keen interest in local history, worked in Edenderry at John Shaw Brown's and was given permission to research its history and kindly provided the following information she found in the factory records:

1897 6 new houses built.

1899 12 new houses commenced.

1899 Royal Ulster Works of Marcus Ward & Co, recently purchased by the Lord Mayor (Mr Otto Jaffe) for £25,000 from liquidator of the company, have been disposed of to John Shaw Brown and Sons Ltd. at an unknown price.

1900 Dining Room commenced.

1901 Factory flooded 12 November.

1902 Gas house built at factory.

1903 12 new houses finished Nos 66-77.

1903 Old yarn shed pulled down for rebuilding.

1905 'Lobby' houses altered Nos 44-57.

1908 'Bunting Bridge' built opposite Front Row in village. Named after factory engineer, Mr Bunting.

1908 Gas lamps in village.

1911 14 new houses commenced.

1935 Electricity installed in village.

1937 March 17-19. Factory flooded to depth of 3 feet.

1937 Bank of 'Dam' burst.

1937 New turbines put in.

1946-1951 Electric plant installed in factory at a cost of £32,024 2s 10d.

1947 Spring. For royal tour of South Africa, J.S.B & Sons Ltd supplied the linen for the royal train.

1972 John Shaw Brown's head office and warehouse, Ulster Works Dublin Road, burnt down as a result of terrorist activity. HQ and warehouse moved out to St Ellen Works.

115 Denis Macneice (see reference 114) provides a similar history for the houses that were known as 'the Cage', stating that they were the first houses built by the factory and comprised eight back-to-back two-storey houses with a hipped roof that spanned the whole complex, constructed in 1866 and demolished in 1929.

116 Nothing now remains of the houses called the Honeycomb which were on the County Antrim bank of the River Lagan on the slope below Longhurst. The Honeycomb and the boat across the Lagan that was operated by ropes are also recalled by Auntie Girlie (references 125, 126).

117 The Duck Walk was also recalled by other local residents. It is thought that this lovely local name refers to the road that connects the Upper Mealough Road with Carryduff, alongside of which is a small stream that feeds into Purdys Burn.

118 Mrs Weir of Milltown kindly provided the following names of places around Edenderry Village: 'The road from Minnowburn to Browns' house was known as Brown's Avenue. The Browns lived in a large house by the river, at the bend of the avenue. About half was along this road, on the side of the road towards the river, by a gate into a big field, were well-grown beech trees close together. At one time this was a heart-shaped plantation, created by Mrs Brown (nee Ellen Kertland). The plantation was known as "The Heart of Trees" or more commonly "Hearty Trees". Just before coming into Edenderry the road goes through a little hill. This was known as the Green Road. Coming out of Edenderry, turning right at the T-junction, part of the road now called Ballynahatty Road was called Hazley's Hill. Hazley, Brown's land steward, lived at the top of the hill. The road going down from the Ballynahatty Road to Ballylesson Road was called the Cut. Further on, at a sharp bend, there is an overgrown lane to the Ballylesson Road, ending at the Masonic Hall. It was known as the Sandy Path, sometimes "Sandy Paith". Near Leverogue crossroads, on the way to Drumbo, a small roadside stream disappears into a patch of woodland and one can hear a small waterfall. This is the Grey Mare's Tail. The local boys used to swim in it, so girls were not allowed near it when boys were there. Across from where Mill Road meets Ballylesson Road there used to be a path down to the High Sluices on the Lagan, but this has been blocked off for some years. There was another way down from Ballylesson Road to the river, where there were cottages and a way to Edenderry. This was Wilgar's Lane. It began from what is now Cameron's nursery. The

Wilgars lived in what is now the nursery building. Across the road was land called the Pound on which there were a couple of cottages.'

119 Annie later explained that the salaries of the weavers depended on which looms they were operating (someone making cloths would get a different wage to someone weaving damask) and on the length of the material they made. The weight of the material woven was checked to see if there was shrinkage or if any cloth was missing.

120 Further information about the last years of the Edenderry factory of John Shaw Brown can be found in the recordings and transcripts of interviews undertaken as part of the 'Living Linen' project that are held at the National Museums of Northern Ireland at Cultra, County Down. See also 'Candida', 'An Irishwoman's Diary', *Irish Times*, 10 March 1980, p. 11.

121 Concerns about the loss of built heritage and risk of overdevelopment at Edenderry have been raised in a number of newspaper articles. See for example, Anon, 'Concerns are raised about plans for new homes in Edenderry', *The Star*, 2 April 2004, p. 19; Anon, 'Countryside protection group calls for a halt to Edenderry development', *The Star*, 7 May 2004, p. 8.

122 Alan Whitsitt, 'A spark of genius', *Belfast Telegraph*, 29 May 1981, p. 14; Bernard Crossland and John S. Moore, *The Lives of Great Engineers of Ulster*, 1, (Belfast: Belfast Industrial Heritage Ltd, 2003), pp. 16-21. Mr McClinton's job as chauffeur to John Brown is recalled in his obituary, see Alfred S. Moore, 'He was the Chauffeur of Ulster's First Motor Car', *Belfast Telegraph*, 5 February 1944, p. 2.

123 See reference 93 for the history of the large estates.

124 The section of the canal near Ballydrain where the lightermen would not stop and other stories of ghosts are mentioned by May Blair, *Once Upon the Lagan: The Story of the Lagan Canal* (Belfast: Blackstaff Press, 1981), pp. 14, 57. The story of the ghost of Haddock is also given by Matthew Neill, *Ecclesia De Drum. Recollections of the parish of Drumbeg diocese of Down* (Drumbeg: Parish of Drumbeg, 1995), pp. 19-20.

125 John Brown's ferry across the Lagan which worked by ropes, is also recalled by Derek Seaton in his stories about Edenderry. The second edition six-inch Ordnance Survey map, Antrim sheet 64 of 1901, shows 'Ferry (private)' and 'Boat House' on the Lagan near Longhurst.

126 Derek Seaton in his stories also referred to the houses known as the Honeycomb, which were situated on the sloping ground between Longhurst and the River Lagan. Two rectangular buildings called 'The Honeycomb' are shown on the second edition six-inch Ordnance Survey map, Antrim, sheet 64 of 1901, and on the 25-inch map of 1901, these buildings can be seen to comprise four or five dwellings. Nothing is known of their history or how they received their name.

127 The history of Sir Thomas and Lady Dixon Park is given by Robert Scott, *A Breath of Fresh Air: The Story of Belfast's Parks* (Belfast: Blackstaff Press, 2000). The history of the rose society is given by Craig Wallace, *And Lovely is the Rose* (Belfast: The Rose Society of Northern Ireland, 2008).

128 The dye works and public house in Drumbeg are described by Eileen Black, 'Drumbeg 1860-1910: A summing up', *Lisburn Historical Journal*, 9 (1995). For further details about the lock house and shoe shop see Anon, 'Down your street. Fresh horses for Dublin stagecoach were kept here', *The Ulster Star*, 27 January 1962, pp. 12-13; Jack Loudan, 'History in stone. The lock-keeper's cottage', *Belfast Telegraph*, 2 February 1963, p. 5.

129 Matthew Neill, *Ecclesia de Drum. Recollections of the parish of Drumbeg diocese of Down* (Drumbeg: Parish of Drumbeg, 1995).

Picture Credits

Page 6: Queen's Quay and River Lagan. BELUM.Y3527. Photograph ©National Museums Northern Ireland. Collection Ulster Museum.

Page 8: Map by Catherine Leinster, think studio.

Page 11: Photograph of Queen's Bridge from up river with docks behind and barge being poled by two men in foreground. HOYFM.WAG.0093. Photograph ©National Museums Northern Ireland. Collection Ulster Museum.

Page 13: The Lagan from Botanic. Reproduced by permission of the Deputy Keeper of the Records and the Public Record Office of Northern Ireland. PRONI LA7/8HF/1P2.

Page 14: Black guillemot. Reproduced by permission of Ronald Surgenor.

Page 17: May's Market. Reproduced by permission of Belfast Telegraph.

Page 18: Rose Ann Smyth and relatives. Image provided by Rose Ann.

Page 21: Billy Shannon at the Gasworks. Reproduced by permission of Billy Shannon and Ray Duncan.

Page 25: Photograph of a John Kelly barge, Belfast Gasworks. Northern Ireland Environment Agency: Built Heritage, 85206. ©Crown copyright. Reproduced with the permission of the Controller of Her Majesty's Stationery Office.

Page 26: The Gasworks and Blackstaff River. Reproduced by permission of Belfast Telegraph.

Page 27: Paddy Lynn, Photograph by the author.

Page 29: St George's Market. Photograph by the author.

Page 31: McConnell Weir and lock. Reproduced by permission of Belfast Telegraph.

Page 33: Photograph of the Gasworks and McConnell Weir. Northern Ireland Environment Agency: Built Heritage, 116546. ©Crown copyright. Reproduced with the permission of the Controller of Her Majesty's Stationery Office.

Page 36: Clock at the Ormeau Bakery. Photograph by the author.

Page 37: Alan Wilson. Reproduced by permission of Belfast Telegraph.

Page 39: The Palm House. Reproduced by permission of Belfast Telegraph.

Page 41: Postcard of Ulster 71.

Page 44: Seal. Reproduced by permission of Ronald Surgenor.

Page 45: The Lagan Weir. Photograph by the author.

Page 47: Dredging the Lagan. Photograph by the author.

Page 48: The Lagan at Stranmillis. Reproduced by permission of the Deputy Keeper of the Records and the Public Record Office of Northern Ireland. PRONI D2334/6/2/1–13.

Page 50: Map by Catherine Leinster, think studio.

Page 52: The first lock at Stranmillis (1850) by James Glen Wilson. Photograph ©National Museums Northern Ireland. Collection Ulster Museum. BELUM.U4970.

Page 55: Stranmillis Weir. Photograph by the author.

Page 56: River Lagan view upstream of first lock, 1888. BELUM.Y4034. Photograph ©National Museums Northern Ireland. Collection Ulster Museum.

Page 58: A bend on the Lagan with rowing boats and people. Photograph ©National Museums Northern Ireland. Collection Ulster Folk & Transport Museum. HOYFM.WAG.093.

Page 61: The Lagan from Stranmillis. Reproduced by permission of the Deputy Keeper of the Records and the Public Record Office of Northern Ireland. PRONI LA7/8HF/1P3.

Page 62: Horse-drawn wagon at the Lagan Vale brickworks. Reproduced by permission of John Hanna.

Page 63: Poling a barge at Stranmillis. Reproduced by permission of the Deputy Keeper of the Records and the Public Record Office of Northern Ireland. PRONI T2850/1/23.

Page 65: Mervyn Patterson. Image provided by Mervyn Patterson.

Page 66: Houses in Laganvale Street. Photograph by the author.

Page 68: Letter heading for the soup works. Reproduced by permission of the Deputy Keeper of the Records and the Public Record Office of Northern Ireland. PRONI FIN 17/1/C/399.

Page 71: Ernie Andrews. Image provided by Ernie Andrews.

Page 73: Annadale allotments. Reproduced by permission of Ernie Andrews.

Page 77: Lagan Vale brickworks and Malone Golf Club. Photograph kindly provided by John Hanna.

Page 78: The old brickworks, Annadale by Elizabeth Holmes Kinnaird fl. 1920-38. Photograph ©National Museums Northern Ireland. Collection Ulster Museum. BELUM.P41.1976.

Page 79: Reproduction of front cover of a catalogue: Anon, *Lagan Vale Brick and Terra Cotta Works Ltd.* (Belfast, 1909). Copy in National Library, Dublin.

Page 81: Former soup works office building. Photograph by the author.

Page 84: The Vulcanite factory. Reproduced by permission of Esler Crawford.

Page 86: Sketch of tanks at Vulcanite. Reproduced by permission of John Johnston.

Page 87: The second lock. Reproduced by permission of Jean McDonnell.

Page 88: Robert and Elisabeth McCurley. Reproduced by permission of Jean McDonnell.

Page 89: Elisabeth McCurley. Reproduced by permission of Jean McDonnell.

Page 91: Mr Byers. Reproduced by permission of Forest Service.

Page 92: George Kilpatrick. Reproduced by permission of Dorothy Kilpatrick and May Blair.

Page 94: Garden and mansion house, Belvoir Park, Newtownbreda, Belfast. With girl sitting by the edge of a stream in the foreground. Royal Irish Academy, Praeger Collection. Shelf Mark Black Crypt 1/3/E. Photo series 18/Box 1. By permission of the Royal Irish Academy ©RIA.

Page 97: George Kilpatrick in his shop. Reproduced by permission of Dorothy Kilpatrick and May Blair.

Page 98: Three of the Kilpatrick children. Reproduced by permission of Dorothy Kilpatrick.

Page 99: Sarah Kilpatrick. Reproduced by permission of Dorothy Kilpatrick.

Page 101: Photograph of the weir near Shaw's Bridge. Reproduced by permission of Ernie Andrews.

Page 103: Photograph of the third lock by the author.

Page 106: Photograph of the lock house and lock by the author.

Page 107: Newforge Ltd. BELUM.Y4059. Photograph ©National Museums Northern Ireland. Collection Ulster Museum.

Page 108: Magazine advertisement for Newforge Fine Foods, 1949, purchased on eBay. No further details known.

Page 110: Newforge Ltd. Sausage production. BELUM.Y4066. Photograph ©National Museums Northern Ireland. Collection Ulster Museum.

Page 111: Photograph of Clement Wilson. Reproduced by permission of J. W. Wilson OBE.

Page 112: Shaw's Bridge, view from left bank of River Lagan above bridge showing barge emerging from under arch and horse being led along towpath. BELUM.Y7657. Photograph ©National Museums Northern Ireland. Collection Ulster Museum.

Page 114: Map by Catherine Leinster, think studio.

Page 117: Shaw's Bridge, view from west bank of River Lagan on south side of bridge looking north-eastwards. BELUM.Y7654. Photograph ©National Museums Northern Ireland. Collection Ulster Museum.

Page 118: Canoes on the Lagan. Photograph by the author.

Page 121: Drum Bridge. Photograph by the author.

Page 122: Edenderry Village and Lagan. Reproduced by permission of the Deputy Keeper of the Records and the Public Record Office of Northern Ireland. PRONI D3670/B/10.

Page 123: The Goat public house. Reproduced by permission of Mrs Weir.

Page 125: Milltown Village. Reproduced by permission of Belfast Telegraph.

Page 126: Thompson's stackyard. Reproduced by permission of the Deputy Keeper of the Records and the Public Record Office of Northern Ireland. PRONI D3670/B/14.

Page 127: Threshing mill. Reproduced by permission of Mrs Merrill Charlton nee Gray.

Page 131: William Glover. Photograph by the author.

Page 133: Buildings at Purdysburn. Photograph by the author.

Page 135: Essie Murray. Photograph by the author.

Page 137: Drawing of estate cottages in Purdysburn Village. Reproduced by permission of John Maddock.

Page 138: Purdysburn House. Reproduced from a photograph provided by Eamon O'Rourke.

Page 140: Edenderry Village. Reproduced by permission of Mrs Weir.

Page 143: Derek Seaton on his electric bike. Reproduced by permission of Sally Seaton.

Page 144: Edenderry Village. Photograph by the author.

Page 145: Susie White, Annie McCarter, Marion McWilliams and Alan McWilliams. Photograph by the author.

Page 147: Susie White at Edenderry factory. Reproduced by permission of Susie White.

Page 150: The fourth lock. Reproduced by permission of Alan McWilliams.

Page 155: Auntie Girlie. Photograph by the author.

Page 156: Mr John Brown at the wheel of his Serpollet car. Reproduced by permission of the Deputy Keeper of the Records and the Public Record Office of Northern Ireland. PRONI D3670/B/1.

Page 159: Lagan ferry at Longhurst. Reproduced by permission of Mrs Weir. (Copy at PRONI D3670/B/1).

Page 160: Auntie Girlie. Reproduced by permission of Eileen Irvine.

Page 161: Eugene Trainor. Image provided by Eugene Trainor.

Page 162: Lion Dance. Reproduced by prermission of Belfast City Council.

Page 163: The rose gardens. Photograph by the author.

Page 166: The Drumbeg dyeworks. Reproduced by permission of Clara Crookshanks.

Page 167: Photograph of the sixth lock, Drum Bridge, provided by Clara Crookshanks.

Page 169: Matt Neill. Photograph by the author.

Page 170: Drum Bridge. Photograph by the author.

Page 171: Drumbeg Church by Niall Timmins. Commissioned by the Forest of Belfast.

Contributors

Albert Allen.
Albert was born in Laganvale Street, the second youngest of five boys. He attended Stranmillis Primary School until the age of 14. Work was hard to find, so he often accompanied his elder brother, a gardener in the Malone area. Casual work sometimes was undertaken caddying on the old Malone Golf Course situated on what we now call Lagan Meadows. Married in 1939, he had two sons. He eventually started his own business in Belfast, working until 1974 when he retired. His interest was gardening, he took great pride in his garden. He loved walking his dog along the Lagan towpath and in Belvoir Park. He is looking forward to his hundredth birthday in 2011. (Contributed by his son, Roy Allen)

Ernie Andrews.
I am in my mid-60s now, retired and still fishing. A few years back I was trout fishing at Lough Money near Downpatrick. It was summer and the lake was low. I was wading and I spotted bricks on the lake bed. I lifted a couple out to see them. They were ANNADALE BRICK Co. ORMEAU ROAD BELFAST. Your past is never far away.

Clara Crookshanks.
I was born on 9 July 1942. I lived in Drumbeg, attended Charlie Memorial School and business school in Lisburn and then my first job was in Barbour Threads. I married my husband Hugh in 1963 and have had two children, Jill and David. We moved away from here for 16 years, to Colin and Lambeg, then back to Drumbeg.

Peter Gallagher.
I was born in Derry in 1966 and I started work as a technician in Queen's heavy structures laboratory in the materials testing section and spent six happy years destroying bricks, concrete and other building products. In 1990, I joined DOE as a civil engineer and in '95 became involved in the design and maintenance of harbours and marine structures. In 1998, myself and three colleagues trained as commercial divers and I then spent several years with the CPD underwater inspection team examining all manner of underwater structures including bridges and harbours. In 2007, I took up the post of river manager at the Lagan Weir.

Auntie Girlie (Susan Irvine nee McClinton).
Auntie Girlie was my much loved aunt – a great cook, baker and knitter, she was as beautiful in spirit as she was in looks. She loved to have people calling in and would welcome all age groups with the warmth of tea and cakes – all home baked. She would always listen to people and was a great counsellor for broken hearts. She loved to sing and could recite every verse of 'The Ancient Mariner' well into her nineties, she was a treasure not to be found today. I will always remember her with love. (Contributed by Eileen McClinton)

William Glover.

I was born on a farm outside Tempo in County Fermanagh, Enniskillen Road, on 4 January 1925. I was the oldest of three in the family. I stayed at school until I was almost 17 and then got an office job for a year. I thought I would make an application to join the police, but my height at the time was about an inch under, they demanded five foot nine and I was only five foot eight. A chum of mine, his father was in Purdysburn and he suggested I consider applying there, and I was accepted.

Patrick Grimes.

I come from County Monaghan and I worked in the shale and brick factory in Kingscourt and when I got married I moved to Belfast. I worked on building sites for two years and then started in Vulcanite in 1969 and stayed until it shut down.

John Johnston.

The house I was born in, just off the Whitewell Road, was blown up in the Blitz and rebuilt. I was born on 11 June 1943. We moved to Greenisland and when I left school I worked as a casual labourer. In 1965, Rothmans opened a cigarette factory in Carrickfergus and I got a job as a crane driver, an overhead crane. In 1986, when the place closed, I went to Vulcanite as a forklift truck driver. The factory closed in '89 though I was asked to stay on, and after it was taken over I worked for ICOPAL as a store man until I retired in 2008.

Peter Johnston.

I was born in 1938 and educated at the Technical High School and later in the Building Department of the College of Technology. In 1959, I was appointed to a post in Belfast Corporation and worked there until 1993 when I was assistant director of Technical Services (Cleansing). I have a keen interest in railway history and other forms of transport and am a member of the Northern Ireland Postcard Club. I am now chairman of McCreath Taylor (NI) Ltd, a company specialising in sales, parts and hire of environmental and municipal equipment in Northern Ireland.

Paddy Lynn.

I was born in the inner city in the 1960s and went to live in the Markets in the 1970s. I worked all my adult life in the area. I have two children, three grandchildren. I am still working in St George's Market, my mother and some other family members still work there too.

Dorothy McBride.

I was born on 11 April 1937 in the third lock house near Shaw's Bridge, the second youngest of 10 children all born there. My dad was lock-keeper. I married in 1956 and raised my family not far from the lock house. I am very pleased the cottage has been restored, it is still my favourite place.

Annie McCarter.

I was born in Ballynahatty at a place called Hillhead. We moved to Milltown and later to Edenderry. I worked at the factory until I was married. Afterwards I was in Grand Parade and in 1992, I moved back to Edenderry, and I am still here.

Jean McDonnell.

I grew up at Belvoir Park and later we moved to Ligoniel and then Stranmillis. Now, I live near the Ormeau Road, not that far really from where I was when I was a child. I had three children, all girls. My passion now is my garden.

Alan and Marion McWilliams.

Marion – I was born in Belfast, we moved to Ballynahatty and then Edenderry. I worked in the wages office at Edenderry factory from around when I was 16 until we had a family. Alan – I was born in Edenderry and have lived here all my life. I worked at the factory from 14 years of age as a tenter until I was about 35, around the time we married. I went to Spillers Foods in Newforge until it closed. Then I worked in Easons distribution in Boucher Road. We are both still living in Edenderry.

Alice and Jackie Murray.

We are here since 1969, 41 years, I actually came from the east of the city, Short Strand area. Jackie came from the Falls Road and when we first got married we lived in Spamount Street, it's where Gallaghers used to be, North Queen Street. Jackie got a chance of this house by the weir through his work, he worked for the old Belfast Corporation looking after the pumping station.

Essie Murray.

I worked in the farm at Purdysburn hospital from 1953 to 1956. Later, into the early 1970s, I worked in the Purdysburn Fever Hospital, first in the laundry then in the sewing room, marking sheets, mending sheets, turning up nurse's uniforms. Since I was married I have lived in Purdysburn Village.

Matthew Neill.

Born in Dunmurry in 1918 and went to two different schools in Dunmurry. From 1924 to 32, I lived at Station House, Brookmount, Magheragall. I came to live in Drumbeg in 1934 and have been here ever since.

Maurice Neill.

I was born on Donegall Road but we moved to Laganvale Street, I spent all my days in Laganvale Street. I started off as a message boy for the Belmont Photo Works. In those days you left a job on a Friday and started another on the Monday and I worked in the Laganvale brickworks and at the Vulcanite, I ended up in Shorts, as a riveter.

Mervyn Patterson.

I was the youngest of six sons that followed my sister and commenced school at the age of five years at Stranmillis Public Elementary School, which was on the Stranmillis Road near where the Ordnance Survey is now. I left school at 14 and entered the Post Office as a boy messenger. Later I worked in a builders' merchants and I applied for a job in the second general revaluation of properties and worked as a referencer and as a temporary valuation assistant. I obtained a qualification and accepted a post with a chartered surveyors and estate agents where, as a complete surprise, I was appointed as a partner in 1965. Later I had my own estate agency business. I am now retired and live on Sharman Road, a few hundred yards as the crow flies from where I was born.

Derek Seaton.

I was born in Edenderry and went to Ballylesson School and worked in Edenderry factory from the age of 14. I lived in different houses in Edenderry. Number 50, 37, then number 16 after Sally and I married. We had two children there but it was a damp house so we went up to number 40 and then we needed a bigger house and went to 34, which had three bedrooms, until 2006, when we moved to Belvoir Park.

Billy Shannon.

I started working in the Gasworks straight after I left Everton School, at the top of the Crumlin Road, at the age of 15. That was in 1960. I worked there until it closed in 1990. At 65, I am now enjoying my retirement but have many happy memories of good times spent in the Gasworks.

Rose Ann Smyth.

I grew up in the Markets, later on the Saintfield Road and now just off on the Ormeau Road. When I was 18, I married Billy Smyth, who had been a boxer and lightweight champion of Ireland.

Eugene Trainor.

I was born on 8 September 1951, the first of twin boys with two older and one younger brother, my sister being the youngest, we were all born with gardening, sport and a love of nature in our veins – a dominant gene inherited from our parents. I was the only one to take up gardening as a career, but my brothers' and sister's gardens are a credit. There were many highs in my life in sport, gardening and wildlife; Sportsman of the Year Greenmount 1969, Plaque of Merit Award World, Rose Convention, Houston 2000, and helping designate Lagan Meadows as a local Nature Reserve, 2008. I retired in May 2010 and am happy to say that my children Melissa, Gareth and Michael are all enjoying careers in education horticulture and sport. Now in 2011, my wife Leila and I have become grandparents to Sara Jane.

Susie White.

I was born in Ballynahatty and, like my sister Annie, lived in Milltown and then Edenderry. I also worked in the factory in Edenderry. I moved to Carryduff around 1960 and later moved back to Edenderry.

Alan Wilson.

I spent my working life with Belfast Parks, starting as apprentice gardener in 1953 and retiring as area manager in 2004. Since then I have worked with my wife in a floristry business and I must say, I don't think I have ever enjoyed myself as much as this.

Lynda Young.

I am from the Ormeau, the top end, Ailesbury Road. I went to St Jude's School, which is no more. It was opposite the Curzon. After school I went to Spence Bryson down Great Victoria Street, but I was only there for three weeks and then out for a month and then I went into the Ormeau Bakery in 1968. I worked there as a packer for 29 years. I now live in Elgin Court at the back of the bakery site.